GATHERING *from the* GRASSLAND

A PLAINS JOURNAL

OTHER BOOKS BY LINDA M. HASSELSTROM

NON-FICTION

The Wheel of the Year: A Writer's Workbook
No Place Like Home: Notes from a Western Life
Between Grass and Sky: Where I Live and Work
Feels Like Far: A Rancher's Life on the Great Plains
Bison: Monarch of the Plains
The Roadside History of South Dakota
Land Circle: Writings Collected from the Land
Going Over East: Reflections of a Woman Rancher
Windbreak: A Woman Rancher on the Northern Plains

POETRY

Dakota: Bones, Grass, Sky
Dirt Songs: A Plains Duet, with Twyla M. Hansen
Bitter Creek Junction
When a Poet Dies
Dakota Bones
Roadkill
Caught by One Wing

COLLECTION EDITOR (with Gaydell Collier and Nancy Curtis)

Leaning into the Wind:
Women Write from the Heart of the West
Woven on the Wind:
Women Write About Friendship in the Sagebrush West
Crazy Woman Creek:
Women Rewrite the American West

EDITOR

Journal of a Mountain Man: James Clyman, 1792-1881

GATHERING *from the* GRASSLAND

A PLAINS JOURNAL

Linda M. Hasselstrom

with a foreword by John T. Price

FIRST PRINTING

10 9 8 7 6 5 4 3 2 1

The Wyoming bucking horse and rider trademark in the High Plains Press logo is federally registered by the State of Wyoming and is licensed for restricted use through the Secretary of State.

Library of Congress Cataloging-in-Publication Data

Names: Hasselstrom, Linda M., author.
Title: Gathering from the grassland : a plains journal /
Linda M. Hasselstrom ; with a foreword by John T. Price, Ph.D.
Description: Glendo, WY : High Plains Press, [2017]
Identifiers: LCCN 2017016135 (print) | LCCN 2017028258 (ebook) |
ISBN 9781937147143 (Ebook) |
ISBN 9781937147129 (hardcover : alk. paper) |
ISBN 9781937147136 (trade paper : alk. paper)
Subjects: LCSH: Hasselstrom, Linda M. | Women ranchers--
South Dakota--Biography. | Natural history--South Dakota. |
Ranch life--South Dakota.
Classification: LCC F656.4 (ebook) | LCC F656.4 .H38A3 2017 (print) |
DDC 978.3/033092 [B] --dc23
LC record available at https://lccn.loc.gov/2017016135

HIGH PLAINS PRESS
403 CASSA ROAD
GLENDO, WYOMING 82213

CATALOG AVAILABLE
WWW.HIGHPLAINSPRESS.COM

for Jerry

"Thousands of us hurl ourselves into cities like nuts into a hopper, and there by grinding and rubbing against one another we lose our natural form and acquire a superficial polish and a little more or less standardized appearance. In the country, the nuts are not subjected to this grinding process."

— ARCHER B. GILFILLAN —
Sheep: Life on the South Dakota Range

Contents

Foreword . . . . *ix*
Author's Preface . . . . *xv*
January 17
February 43
March. 65
April 85
May 111
June. 137
July 157
August. 183
September 207
October 239
November. 263
December. 289
Acknowledgments *311*
Bibliographic Essay *312*
Further Reading *318*
About the Author *320*

Foreword:

Grassland Inheritance

NEAR THE OPENING of this remarkable book, Linda Hasselstrom reflects that ranchers "are the children and grandchildren, and probably descendants back to the cave dwellers, of people who worked as long as they had light to see, and we still judge our days by how much we accomplish."

I would like to speak to that accomplishment.

I first met Linda in June of 1994, at a café in Rapid City. I was a 27-year-old grad student at the University of Iowa, working on a dissertation about grasslands nonfiction writers, which would later become my first book, *Not Just Any Land: A Personal and Literary Journey into the American Grasslands*. A few months prior, I had written Linda and a handful of other regional authors to request interviews, and she generously agreed to meet with me—not always the case with busy writers being approached by overly-enthusiastic graduate students. Little did I know, however, as we began that conversation about her life and writing, how it would impact the trajectory of my own.

Like so many born and raised in the American grasslands (in my case western Iowa) I had longed to leave for places I believed were culturally and environmentally more interesting. Although my close-knit, intergenerational family was proud of its deep Midwestern roots, my adolescence was significantly affected by the struggles of our community during the 1980s Farm Crisis, when the surrounding fields—the symbol of the Arcadian ideal so many outsiders thought we were living—became, for me, synonymous with despair and displacement.

At the University of Iowa, I studied literature and writing, and became enthralled with the idea of creating books of my own. It went without saying that success in that arena would require leaving home. After all, most of the regional "greats" I had read and studied—among them Cather, Twain, and

Hamlin Garland—had spent their rural childhoods here, but as adults moved away to fulfill their literary ambitions. Their reasons for leaving were in part practical, seeking closer proximity to research libraries and the coastal centers of literary production, but they were also cultural and ecological. Some, like Garland, spoke of being assigned primarily Eastern, cosmopolitan authors in his rural school, which led him to worship them like "demigods." Others, like Mari Sandoz, recalled local and familial denigration of literary "work," as well as the paradoxical view of escape from rural life as both betrayal and success. Other writers addressed what contemporary Indiana essayist Scott Russell Sanders describes as a residual Puritanism that sought to control and repress the "promiscuous, sensual, earthy" qualities of the human imagination. The irony was that many of the writers who grew up in the rural grasslands of the 19th and early 20th centuries were inspired by those same "earthy" qualities in their daily work on farms and ranches, as well as childhood adventures in the diminishing wild prairies, which they later wrote about from a distance. The destruction of the native grasslands themselves—the greatest ecological loss in American history—also figured prominently in the memories of some ex-pat Midwestern authors and may have contributed to their sense of exile. The "homeland" they knew as children, including its wild beauty, had literally been plowed under.

For many of us who came later, we hardly knew it at all.

That all began to change for me the summer before I met Linda, during the great Iowa floods of 1993. My wife and I had moved to the small town of Belle Plaine, closer to her new teaching job, and I was commuting to Iowa City to finish graduate school. The natural beauty I witnessed during those morning drives—the prairie flowers and grasses erupting in the unmown ditches, the diverse bird-life inhabiting flooded crop-fields—gave me a peek at a lingering, wilder Iowa I had never truly noticed or appreciated. I found myself inspired to learn more about regional natural history, bought field guides to help me identify native flora and fauna, and sketched out some of my first nature essays. Equally important, the destruction wrought by the floods had simultaneously given birth to a new, albeit tentative, question inside me: What would it mean to commit to living in and writing about my home?

To truly respond to that question, I needed more than natural history books and field guides and topographical maps. I needed a map of words, first-hand accounts by contemporary authors who had made a personal commitment to the Midwest and Great Plains, to its natural and human communities. I needed the living example of writers who embodied an alternative version of regional writing, perhaps best defined by Kentuckian Wendell Berry as "*local life aware of itself.* It would tend to substitute for the myths and stereotypes of a region a particular knowledge of the life of the *place* one lives in and intends to *continue* to live in. It pertains to living as much as to writing, and it pertains to living *before* it pertains to writing."

Enter Linda Hasselstrom. I accidentally discovered her books during the same summer as the floods, on the shelves of (appropriately) Prairie Lights bookstore in Iowa City. Despite a limited graduate student budget, I purchased all three volumes—*Windbreak: A Woman Rancher on the Northern Plains* (1987), *Going Over East: Reflections of a Woman Rancher* (1987), and *Land Circle: Writings Collected from the Land* (1991)—and over the next few months devoured them, covering the margins with notes and questions. More than any other grasslands author I had read, her writing emerged from the joys and challenges of a lifetime relationship to one piece of earth—that one-mile radius of buffalo grass prairie in western South Dakota that she referred to as "the circle of [her] world." Over the course of those three books, which described her life from childhood to middle age, I was introduced to the daily rituals of ranch life—the beautiful but bloody ritual of spring calving, the freezing winter mornings, the threat of fire, the intensity of sunsets over the grasses. As I struggled to find balance in my relationship to home, I learned from her own efforts to do so, whether between writing and ranching, between her environmentalism and her family's economic dependency on the land, or between the expectations others had for her as a citizen, a daughter and a wife. I grieved the death of her beloved husband, George, and shared her admiration for her father, the man who adopted Linda at age nine and first brought her and her mother to that land, indelibly linking her to the generations of ranchers in his family. Her position in that place seemed secure, even as she moved toward an uncertain future.

"What will I inherit?" she asks in *Going Over East.* "Perhaps nothing that I don't already have: the knowledge and love of the land I have gained from my father and from my own absorption in this unique world."

The issue of inheritance took on new complexity, however, during our conversation in that Rapid City café in 1994, when Linda discussed the fact that she was no longer living on the ranch, but in Cheyenne. Her father had become increasingly erratic and verbally abusive in his later years, likely due to senile dementia and a series of undiagnosed strokes. After George's death, he demanded that Linda choose between writing and ranching. When she refused, he forced her to leave the ranch and disinherited her. She would write more about that painful experience in *Feels Like Far: A Rancher's Life on the Great Plains* (1999). Eventually, when her mother's health began to fail, management of the ranch returned to Linda, but at the time of our conversation she was renting the houses and the land to strangers. She hadn't visited the pasturelands since her father's death. She had become, in her words, "my own worst nightmare—an absentee landlord."

At the end of the interview, Linda surprised me by inviting me to tour the ranch with her the next day. In retrospect, this generosity shouldn't have surprised, since she had been inviting so many to visit that land in her books, to appreciate its beauty and significance through her eyes. Like Cather's Red Cloud or the Sandoz farm in western Nebraska, it had become a literary landscape. Indeed, when we arrived the next day, many of its major characteristics were familiar. But as she pointed out small, specific spots—a hill, a hidden valley, a tree—that had meant something to her personally, or small changes (to me) that visibly upset her, what had been abstract in her books became real. For the first time in my life, I was witnessing what it meant for one writer to completely identify with one spot of earth. It became clear that Linda's relationship with that ranch was a form of committed love as real as any other, as worthy of joy and sacrifice and perseverance as any other, and as full of emotional risk, as she faced the very real possibility of losing it forever. Even so, when we left the ranch that afternoon, I could tell that she was determined to find her way back.

This wonderful new book, *Gathering from the Grassland: A Plains Journal,* is Linda's multi-layered story of return. In it, she reflects on her return to the ranch, and her renewed commitment to living there as a writer and teacher—the very passions that led to her exile—now sharing it with other women writers and artists as a retreat. It also marks a return to the journal form, which she originally used in *Windbreak*. It is a literary form that demands a degree of discipline reminiscent of her work as a rancher, but which also immerses readers in the kind of attentive wonder required to see in individual, daily life the multitudinous ways we are connected to others and to the larger sweep of time. This includes connections to the natural world immediately around her, where blooming wildflowers and bison dung and the paw prints of coyote and mountain lion are, as she puts it, their own kind of journal, full of mystery and wisdom for those willing to learn. It also includes people, past and present, such as her long-time partner, Jerry, who have influenced the person Linda has become. Of particular significance are her now deceased parents, whose journals provide Linda with new insights into the difficulties that separated her from them, as well as into the values and habits of mind that still link them, revealing opportunities for understanding.

Most importantly, all of this takes place on the ancestral land that continues to define her, feeding both body and spirit.

From a literary perspective, *Gathering from the Grassland* provides further confirmation of Linda Hasselstrom's status as one of the most important writers in the Plains canon, a status increasingly recognized by writers and scholars in the field. Susan Naramore Maher, for instance, in her recent book *Deep Map Country: Literary Cartography of the Great Plains*, states that Linda Hasselstrom's work is representative of "deep map writing from the Plains" which "engages fully with the material and spiritual contexts of this magnificent if stressed landscape and in doing so gives voice to the accumulated lessons of place." For Maher, the urgency of that work is clear: "We will need to unfold and study the many maps of spiritual geography to avert further diminishment of the grasslands, to face facts, to regard the land's soil and soul, and embrace the living matrices of the North American Plains."

I couldn't agree more with Maher's assessment, but when it comes to the accomplishment of which Linda speaks in my opening quote, it is her model for living that especially inspires me. Once again, her wisdom has found me when I am most in need of it. I am now 50, the same age she was when we first met in that café in Rapid City, and no longer deciding whether or not to commit to place as a writer. Instead, I am looking back on decades within that commitment, having settled only a few hours from where I was born, where my wife and I have raised our three boys—the seventh generation in my family to live in western Iowa—and where I have written books about my own complex relationship to the natural and human communities of home. Linda Hasselstrom's writing, her personal example, helped guide me toward the choice of this life, and continues to do so. This includes her timely reminder, in the midst of the worries of my adult life, to rely on the prairie as "the path to enlightenment and the enlightenment itself" and to every day throw myself "into noticing the red-winged blackbirds, the greening grass, and the signs of returning hope."

But the true accomplishment of this book, of all of Linda's writing, is that its transformative power isn't limited to those who happen to be writers or happen to have an affinity for the prairies and plains. It is available to anyone who has at once embraced and wrestled with the responsibilities of family, work and community. Anyone who has ever thought themselves lost, or disowned, and in need of a home. Anyone who has longed for the sweet warmth of solitude, and an appreciation for the lyrical dignity of aging. Anyone who has sought beauty in the midst of brokenness, and faced the challenge of moving beyond grief and betrayal to accept the affection of another. Anyone who has lost one kind of faith, and found another by stepping outside to "see nothing that is not holy." Her words teach us to notice the extraordinary in the ordinary, and to be attentive to each miraculous hour that remains to us on Earth. Out of love for this life and for this land, wherever we may be.

This is our inheritance from Linda Hasselstrom, and it belongs to all.

JOHN T. PRICE, PH.D.
DIRECTOR OF CREATIVE NONFICTION WRITING
UNIVERSITY OF NEBRASKA–OMAHA
AUTHOR: *Daddy Long Legs: The Natural Education of a Father*
EDITOR: *The Tallgrass Prairie Reader*

Author's Preface

In 1987, my first book, a diary representing a year on my family's small western South Dakota ranch, was published by Barn Owl Books. *Windbreak: A Woman Rancher on the Northern Plains* received positive reviews. *Ms. Magazine* said, "Her account . . . will leave readers amazed, exhilarated," and the *New York Times Book Review* praised "the fine small beauties of her strenuous life."

Hundreds of readers wrote to me with thanks for letting them see ranch life, or offered to work as hired hands. Not long ago, twenty-eight years after the book had been published, a rancher from another part of the state drove into the yard to meet me and to see the place for himself after reading the book.

"A windbreak is a precious thing," I wrote then. "We all need a windbreak." The shelter of juniper trees with its understory of buffaloberry and plum bushes north of my home on the ranch remains. After three decades, the trees are tall and sturdy, the bushes thick with thorns and fruit. The barrier protects the house and abundant wildlife.

I still keep a daily journal, gathering the events of my life into true stories of the grasslands. Everything in the days and year that follows really happened but the book also encompasses many days and several years.

Once I acknowledged that this journal would be published, I included explanations and background information that would not appear if I were writing only for myself. I also tried to write honestly, ignoring the reader looking over my shoulder. I cannot judge my success.

Settled back into Windbreak House in the home I built with my second husband, George R. Snell, I live with my life partner, Jerry Ellerman. I write, read, and try to live responsibly in my community of people, wildlife, and plant life.

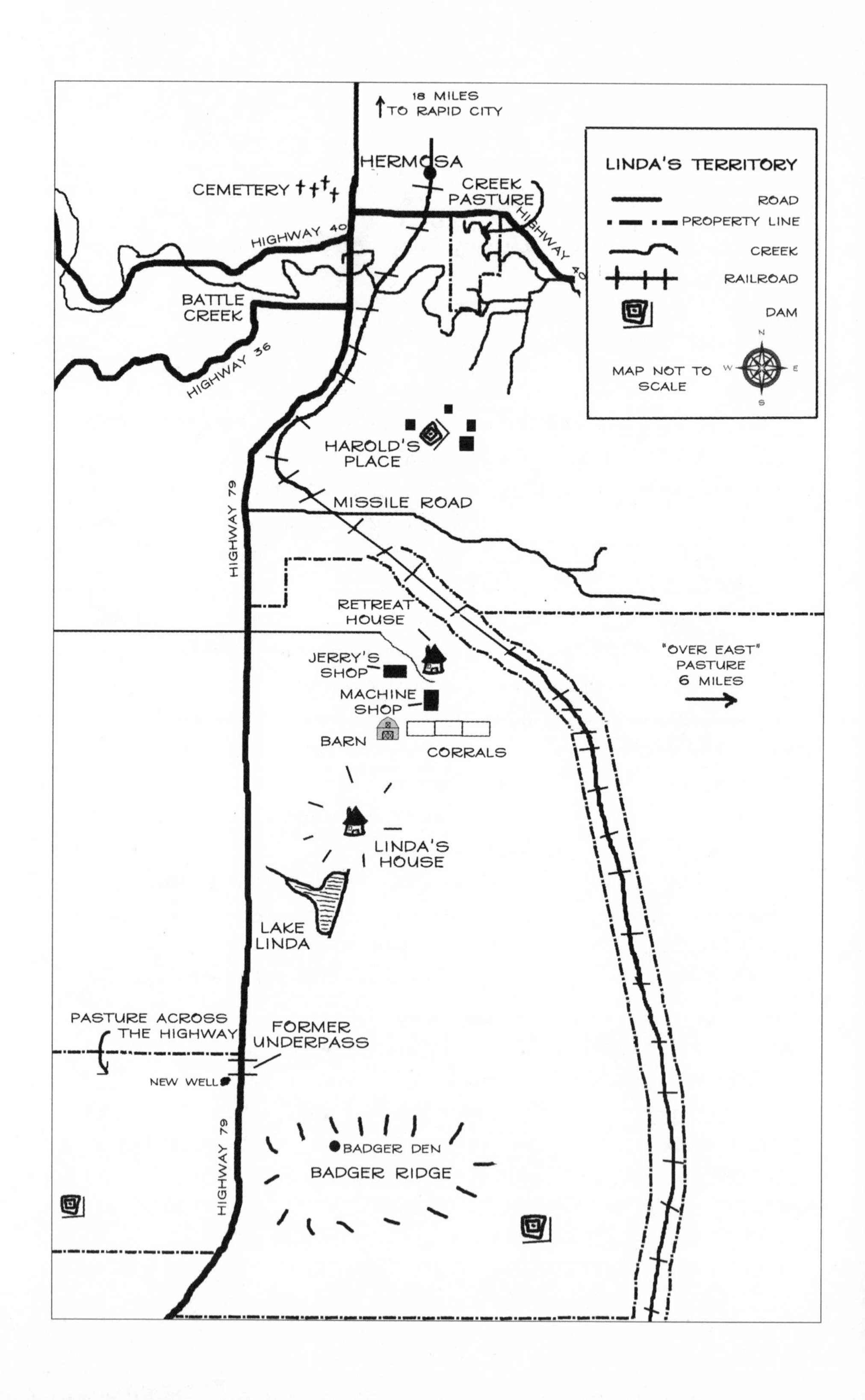

18 MILES
TO RAPID CITY
LINDA'S TERRITORY
ROAD
PROPERTY LINE
CREEK
RAILROAD
DAM
MAP NOT TO SCALE
N
W
E
S
HERMOSA
CEMETERY
CREEK PASTURE
HIGHWAY 40
HIGHWAY 40
BATTLE CREEK
HIGHWAY 36
HAROLD'S PLACE
HIGHWAY 79
MISSILE ROAD
RETREAT HOUSE
"OVER EAST" PASTURE 6 MILES
JERRY'S SHOP
MACHINE SHOP
BARN
CORRALS
LINDA'S HOUSE
LAKE LINDA
PASTURE ACROSS THE HIGHWAY
FORMER UNDERPASS
NEW WELL
BADGER DEN
BADGER RIDGE
HIGHWAY 79

January

January 1. For our first breakfast of the New Year, Jerry pounded antelope chops thin, dipped them in flour seasoned with salt and red pepper, then fried them in olive oil to a crisp brown.

My mouth waters writing this; the chops were so tender we cut them with a fork. As we savored the rich wild essence of antelope, we watched the rest of the herd climb up the ridge to the south, almost invisible in the snowy grass. My field guides say the Latin name is *Antilocapra americana;* none use the term "antelope" but that is how we generally refer to them. The first scouts topped the hill, black horns slicing the air as they looked for trouble. Then the tawny herd flowed along behind them and vanished over the horizon. The grassland provided a feast for our eyes as well as our bellies, and every detail was part of nourishing and protecting this acreage I own.

Once we finished walking the dogs, my life partner, Jerry, went to his shop to create something of metal or wood, and I went to my computer to build with words. He'll be back for lunch, when we'll exchange news, read the paper, and nap before going back to our respective jobs.

Now seated at the computer, I recall our morning walk in the yard of the ranch where I grew up, wrapped in a freezing fog that slowly whitened the branches of the cedar trees. Jerry walked beside me, and two West Highland White Terriers, Toby and Cosmo, chased rabbits somewhere ahead. I inhaled frosty fog that smelled of icy grass and exhaled steam.

My job is words. Self-created job. Doesn't pay much. Stumbling through the world looking for clues to my own existence, I read a lot, and write. I also look at the grass, the trees, and the fog. I listen to the birds and to the voices inside my skull. I taste grass and bee balm and sage and dust in the wind, all of them part of the flavor of the beef grazing on the horizon.

The question of every day, and especially of the first day of each year, is "What's next?" What follows on these pages is a year of exploration, a year of questions followed—sometimes—by answers.

The last time I asked that question aloud, "What's next?" I was speaking with a dying friend. She said, "I'm going to die as quickly and efficiently as possible so as to cause my children the least possible trouble." And she did.

That's the kind of resolution I admire. I'm not planning to die just yet, but I want to write this efficiently and clearly to show what life is like on my ranch today. My first book, a year's diary, was published nearly thirty years ago; many of the people who shared that year with me are dead. I will look back as I look forward.

The fog oozes through the windbreak trees north of the house and flows down the gravel road, subtly as days pass before we realize our own lives are changing.

Windbreak House, my home, stands on a hilltop overlooking the ranch house my parents, John and Mildred Hasselstrom, built and where I grew up. We have a good view of the ranch buildings, various sheds, garages, a machine shop, and the corrals my father worked all his life to keep in repair. Cattle still graze the surrounding land, but they belong now to a neighbor, Rick Fox, who leases the grass from me.

Today my hilltop house is surrounded by fog; I can't see the highway a quarter mile away. The fog is real, but it's also symbolic of what it's like to write about life while living it.

As we sit down to supper, we will say, "A good day. We got a lot done." No matter that we don't run cattle, that Jerry is retired, that my job of writing is entirely self-created. We are the children and grandchildren, and probably descendants back to the cave dwellers, of people who worked as long as they had light to see, and we still judge our days by how much we accomplish.

These predictable lives, filled with work we have chosen, seem like luxury to us; some might call them boring. We have lived through the maniacal insanity of youth filled with demands from others, what one of my friends calls the "drama of the young." Now we answer mostly to ourselves and to each other, with respect.

Though I say that I am a full-time writer now, for much of each day I am not a writer. Though I spend each day writing essays, poems, and this journal, I also live a normal life of shopping for groceries, cleaning toilets, cooking meals, going to meetings, and wondering how I can be so busy when I don't have a job. I volunteer for baking and kitchen service to a local history group. Some days I'm most inspired when meeting the challenge of cooking a healthy meal or collecting dried grasses from the hillside to make a bouquet.

As another exercise in creativity, I've turned my parents' ranch house into Windbreak House Writing Retreats. There I keep my library and conduct writing retreats, as well as working online with writers, most of whom I have never met. In each case, I write long comments to help each person write more successfully. These jobs are all important parts of my life; they are not writing, but part of the bargain I have made with the world in order to lead a satisfying life defined by my own choices.

I once left here assuming I'd never return, but I'm back on this land where my writing life began, enthusiastically becoming involved in the surrounding community. I expect to end my life here, or, perhaps more likely, in some kind of assisted living in this region. Because I am cursed with the Gene of Organization, I'm trying to be methodical in looking toward the winter of my life.

I am wary of journals, even, or perhaps especially, my own. When I read the journals I wrote in the two months before my husband George's death, for example, I am overcome with fury at my own inability to see what was happening to us.

So, if those journals ever became public, how could a stranger possibly understand what I was feeling through what I wrote? Yet for a confused stranger to know how bewildered I was at that time might help another person survive.

Isak Dinesen said, "All sorrows can be borne, if you put them into a story." John Steinbeck once wrote, "If a story is not about the hearer he [or she] will not listen. . . . The strange and foreign is not interesting—only the deeply personal and familiar."

So we are all looking for ourselves in the stories we read, trying to discover if we are living our lives correctly according to some universal standard. Or perhaps searching for where we went wrong and how we might correct our own trajectory.

January 2. Whenever I go to my study, I realize my life has been full of people who documented their lives in one way or another. My father apparently began keeping his record when he married my mother, recording mostly temperatures and ranching information. I've begun reading all of my father's journals that I could find. They begin about 1956 and continue through his death in 1992. From my reading chair in the living room, I can look east, down to the retreat house, the barn, the corrals where he spent much of his working life

From his journals I have copied the high and low temperatures he recorded for every day into another set of journals, creating a record of the weather here for thirty or forty years. Studying these statistics, I thought, might help me decide if climate change is occurring in this one small part of the world. I planned this project to absorb my attention at times when reading or my own writing was not keeping me occupied. Each day I add the high and low temperature and a few words describing the weather.

As a person who has a difficult time throwing away books or papers, it is my misfortune to be descended from many similar packrats. Yet it is such records that have provided insights to historians throughout the ages. Where such papers were not found, often the historians lament the lack of evidence for the way ordinary people conducted their lives.

Since my father was the biggest influence on me, perhaps it's fitting for me to start this year studying his journals. Or maybe I am making this job seem important so I can feel as if my father's records, and therefore my own journals, are part of an historical record, creations more important than self-centered jottings.

At first, he wrote sparse notes, mostly the temperatures. On this date in 1956, he wrote only that it was 36 degrees, presumably the high temperature for the day. Later, he began to add the low temperature, often recorded when he got up at 6 A.M. or earlier. Gradually, he included a few words about the weather or the work he was doing. On this date in 1968, for example, he wrote *"15°. Looks mean."*

My project of copying only the high and low temperatures quickly became tedious because I resisted reading the rest of his entries. Gradually I began to read more, realizing that my father didn't reflect on his life much, just reported information that would help him in the work of raising cattle in this particular spot. Soon I began to recall the flow of those days, many of which I shared, the way the work ran on from one day into another. Like most people who work for themselves, my father had considerable freedom. He didn't have to get up or go to bed at a particular time in order to punch a time clock. He didn't have to dress to suit anyone else's rules. His bosses were the cows that formed the basis of our livelihood: their health and productivity was his primary interest.

January 3. Sunrise is a clear green sky behind the black line of the hills. Low-lying clouds turn gray and reddish-gold in alternating bands.

A half-mile southeast of the house, two oval shapes are perched close to the top of an old cottonwood. With the binoculars, I can see they are two mature bald eagles. For an hour, while we play Rummykub at the dining room table and then walk the dogs down to the ranch buildings and back, the two eagles remain. They are surely aware of us, but confident we pose no threat to them. Naturally I am looking somewhere else when they fly off.

When my parents took one of their rare vacations, my father didn't take his journal with him. I realize only now that this means he saw those documents as not about *him* but about the ranch, the place. After I divorced my first husband, my father and mother began taking a winter vacation in Texas; he'd leave his journal for me so he'd have a record of the winter temperatures and the work I'd done.

A rancher often speaks as if his ranch, his place, is himself, as in "I ain't had any rain on me since May." So my father's journal was about this particular segment of the great central grasslands. When he was away from here, his actions and observations were of no interest to him.

I'm still copying the temperatures for all the years he recorded them, adding them to the temperature notebooks I am keeping. When I talk about this project, people always ask if the temperatures have gotten steadily higher. No; throughout most of that time the fluctuations have been slight and brief, so I can't say that the climate is definitely getting warmer; all I can deduce is that the weather isn't getting any more predictable.

As I read my father's journals, I sometimes read my own records from the same time. Though I lived what I wrote, I am still astonished at all the work I was doing in the 1970s and early 1980s. Besides gardening and housekeeping, I rode almost daily, moving the cattle here and there, dashed to town to haul supplies, and wrote in the evenings or when rain kept us out of the pastures. As I read, I absently rub one or the other of my knees, snapping with pain that is probably the result of all that hard work. Like my wrinkled face, these are the knees I have earned.

If I had not worked so hard at ranching, would I be a better, or better-known, writer? If I had not worked so hard at writing, would I be a better rancher?

January 4. Just at sunrise, we saw four huge canines, dark brown and at least as big as German shepherds, trotting down the draw. One stared at the house, but they didn't pause and sniff and ramble like coyotes, just passed like warriors on the march, packed closely together, staring toward the house. Later in the day we found their tracks in snow six inches deep, marching in a straight line past a rock outcropping and into the neighbor's pasture, headed north.

Online research told me that coyote tracks zigzag, but wolves walk in a straight line. When I reported to a regional Game, Fish and Parks officer that we might have seen wolves, he was not enthusiastic but agreed with my reasoning. "Probably just passing through," he said in a reassuring tone. Later standing on the deck, I feel the darkness pressing against me. I know wolves may be watching.

Still, we have known for years that mountain lions prowl this valley. I was probably within a few feet of one while checking calving heifers years ago, and we've seen and measured their tracks in the snow. Somehow the cats seem less real than the wolves tonight—because we have seen only the tracks of the lions.

Tracks. Like the writing in journals. Ephemeral. Snow-borne tracks blow away with the wind. Scribbled lives on paper disintegrate, mold, and turn to dust.

Heading to the kitchen to check the roast I put in the oven at 8 A.M. for lunch—beef nourished with grass from my pastures—it occurs to me that my father's journals are only a small part of my collection of papers my ancestors have provided. Keepers of records, all of them. Besides my mother's diaries, I have letters, photographs, invoices, and other papers by the hundreds. I spend the afternoon sorting, categorizing, and arranging boxes full of paper.

Since I am evaluating my life, I have to ask myself if my father's journals are important to anyone, even me. His life is over; he did not write self-analysis in his daily pages, only the barest facts of his life as a western South Dakota rancher.

I don't want to face that question now.

January 5. Snow melts in the yard first, so the ground is brown and beaten where we've walked. Our fingertips and lips crack like the evaporating ice on the pond below the house, leaving brown weeds sticking up to rattle in the wind. A neighbor called the pond "Lake Linda," and we've adopted the term.

Today I studied my favorite photograph of my father's family, most of whom I never met because they were dead before my father married my mother and adopted me. The family stands in front of the gray, unpainted porch of the ranch house. The women wear loose housedresses; the men stand broad-shouldered in overalls. Clearly it is summer, and probably they have been working.

My father appears to be about ten, placing the time around 1919. He stands at the far right, barely in the picture, a couple of feet to his father's left. He is wearing an oversized shirt buttoned at the neck and too-large overalls rolled up at the bottoms. His broad-brimmed hat is pulled down so only his nose and mouth show. His father, Charles, also wears coveralls rolled at the bottom; he's not wearing a hat. Did he put it on John's head at the last minute as a joke? John's hands are stuffed in his pockets, and he stands a little sideways, his expression grim. He'd already had rheumatic fever that kept him in bed for months, stunted his growth, and left one arm shorter than the other; his left hand is stuffed into his overall pocket, perhaps to hide what he considered as a major defect. He referred to himself bitterly as a "cripple." In this picture, he stands so alone that he seems truly outcast.

Suddenly I see him as he described himself: as a misfit in this family of big, healthy hard-working people. Because he was crippled, he said, they decided to send him to college, considering him unfit for ranch work. His youngest sister also got to attend college, because she was a girl, and by the time she was ready they probably had more money. He got an education and worked in offices in South Dakota and California until his next older brother was killed in World War II. Then he came back, divided the family ranch with his brother, and until the day he died at age 92, he worked hard at ranching, as if determined to prove his family wrong every day of his life. His wounds were not visible as scars, but in his behavior.

Perhaps some of the anger I often thought was directed at me was really at his own circumstances, his own blood. From whatever he felt and believed, he created a legacy for me: I rarely pass a day without thinking of some of the precepts he instilled in me or some of the criticism of me that marred our last days together.

I'm still trying to prove myself to him, sixty years after he adopted me and twenty after his death. Now the proof becomes serious: I own the ranch to which he devoted his life, and my actions will decide who owns it next.

January 6. 10 below zero with a ferocious wind chill. Twenty sage grouse scratch in the junipers as if it were the dust of a chicken yard, clucking and pecking and flinging snow over themselves. No wonder the settlers called them "prairie chickens."

When George and I built this nondescript Little House on the Prairie, we planted a windbreak of juniper trees and bushes like buffaloberry and chokecherry. I watered the trees a little the first summer and then declared that they must live on what nature provided. They have thrived. The junipers are ten to fifteen feet tall, with thick branches close to the ground. Sometimes the dogs chase a rabbit under a tree, and we can see it hunkered among the branches, probably safe even from a coyote. A human who crept close to the trunk might survive a blizzard. The windbreak shelters the grouse and smaller birds, and probably also the predators who hunt the smaller critters.

As we walk down to the ranch buildings, the air is filled with fine sparkly glitter, bits of ice or frost that must be blowing up from the grass or freezing as it falls to sparkle on our eyelashes, our hair, and our stocking caps.

Extreme cold reminds us who and where we are; I fear we mention it too often to friends in warmer climates. South Dakotans also like to say, "It could have been worse," taking perverse pride in what we have survived.

Walking through the ranch yard, I realize that the Hasselstrom family has been ranching on this shortgrass prairie for more than a hundred years. Karl Hasselstrom came from Sweden, Anglicized his name to "Charles," married a widow with three children, and produced five of his own. From such beginnings have come ranching dynasties, but the Hasselstroms, says one neighbor, were too busy working to reproduce.

One of my aunts frequently mentioned that I am "not a real Hasselstrom" because I was not born with the name. Ironically, I am the survivor, the only one of the family name left in this area. In this seventh decade of my life, it's time to make plans for what will happen to the family ranch that cobbler Karl Hasselstrom began by taking up a homestead here.

Looking toward the future of the land, I've decided, requires that I spend some time looking back at the history of my family, studying the documents they left behind. As a writer, I am perhaps more comfortable with written records than with conversation, and that surely has been true of this family of mine.

January 7. I open my journal to this date in 1986 and read, *"It's so hard to get up this morning. It seems logical to stay in bed and so illogical that a writer my age, with my credits and professionalism, should have to struggle out of bed in a cold house at 7 and* HOPE *to get 1½ hours of work before time to go out and spend the rest of the day feeding cattle."* I struggled to find writing time when George and I were doing the ranch work. We would finish feeding by two in the afternoon, I'd fix lunch, and then we'd collapse for a brief rest, aching. I would then try to write a little before we went out for the evening chores.

The journal entry reminds me how exhausted we were during that year, and then suddenly he was terminally ill and then he was dead. In my darker moments, I say that we worked him to death on this ranch, though I know that the tumor that killed him was already in his spine when we met; he would have died from it no matter what work he was doing. But instead of all the labor we did, we might have been enjoying our lives more.

Now, thirty years later, I am more likely to suggest to my partner that we break from our work to do something enjoyable: watch a movie, have lunch in town, go for a drive or a walk with the dogs.

On this day in 1957, a Monday, my mother wrote, *"A lovely washday. . . . and I* FELT LIKE WORKING." What a good reminder that I should concentrate on the positive things my parents wrote! I'm so invigorated that I take a break from my writing to mop the kitchen floor.

At nearly four in the afternoon, I take the dogs for a walk because the temperature is a balmy 31 degrees. On the west side of the hill I look around: the sky is a flat and uniform gray. The ground is mottled white and tan with the light snow that has blown away the past few days. The sun will set soon but the clouds are so thick that we are likely to see only a faint glow. Altogether, the scene is not cheerful.

Then I look down. Long bunches of redtop grass and little bluestem bent by the recent snow are glowing vivid pink with light reflected down through the clouds. My eyes see only gray in the sky, but the facets of the grass collect the light to shine it back at me. Brown spiky heads of echinacea contrast with the shimmer of big bluestem and turkey foot. The world's color is beneath our feet and paws.

The dogs scent a rabbit and gallop toward the circle of cedars where Jerry placed a sandstone bench. While they happily yip and flail through the grass, I sit to review the lesson: sometimes it's better not to look at the big picture, the long view. I must remember to look at the nearby.

January 8. Sometimes I skip several days of copying my father's temperature notes, keeping the journals close by my reading chair but ignoring them because the writing makes my hand cramp.

Lately I've begun to see revelations in his journals, things I'd forgotten or never known. When my old dog MacDuff's epilepsy became so severe medication didn't help, my father killed the dog and buried him in the pasture, marking the spot with a large stone. When I came back from college, he told me, quietly, what he had done and why: this was part of my education in how the real world works. In doing those hard tasks, he was also teaching me to do things that are difficult but right, a lesson I've remembered, though I snapped at him at the time and later cried.

January 9. As the sky began to lighten to sunrise, I looked south and saw a line of cows on the ridgetop, silhouetted like sculptures. While we were having coffee in the bedroom, we saw a coyote trotting down the hill south of the house. Our windows framed the scene, a wide white stage where we could see the coyote for a half hour. Every few feet the coyote, probably a female, since several females band together to support the puppies produced by the alpha female, would stop to look alertly around. Then she'd pause, head tilted downward, listening for a mouse or vole scrambling under the snow. When she pounced, she kept her front paws together, ears pricked forward. Sometimes she'd toss something in the air and swallow it before moving on a few feet.

January 10. 30 degrees at 5 A.M. A fingernail moon is waning over the south ridge. Jerry says at sunrise that he probably will grill steaks tonight; the forecast is for temperatures in the 40s and 50s.

I've just learned that Wes Harrison, one of my dad's oldest friends, a long-time neighbor and one of the last of the old-time ranchers, died a day or two ago.

After George died, one day when the temperature was forty below zero, I heard a snowmobile coming through the pastures from the east. I stood at the window with binoculars, cussing the @#$@$@% trespassing snowmobiler, who rode his noisy machine right up to my porch. When I stepped outside, I realized it was Wes. He'd been widowed himself by then and had just been "riding around," he said, and came to see if I was okay. He wouldn't even come in for coffee, but he reminded me to call him if I needed help.

Waking up this morning I had a sudden vision of him and my dad meeting wherever it is that good old ranchers go.

"Well, howdy, John!"

My father nods. "How are you, Wes?"

"Not bad, you?"

"Pretty good."

They nod at each other, understanding and glad to see each other even though it means they're both dead. "Good grass this year?" my dad asks.

"I've never seen it any better," Wes says, grinning. "Why that pasture of mine over by your east fence. . . ." They walk off together, two lanky old men in loose-fitting jeans.

I cried when I heard the news of his death and am crying again writing this. But because, as some writer surely said, to a writer everything is relevant, I have to ask myself questions while I sniffle. Am I crying for my own lost past, for what these men meant to me as I was growing up, the stability they represented? Or for the loss to the community?

Wes was a diamond of a man, a wealth of kindness, and of knowledge about cattle and people. His folks, Lee and Lillian, owned a ranch to the east of ours, bordering on our easternmost pasture. Whenever we got in trouble in our east pasture—ran out of gas, got stuck, or were caught in a blizzard—Harrisons were nearby and ready to help. Often when we moved cattle over east, we used only horses. We couldn't trust my mother to follow us with the pickup; she invariably got lost. So if we got caught in a rain or hail storm, we'd ride down to Harrisons. They'd feed us as our clothes and hair dried and we warmed up; I'd listen as the men talked. Lillian always seemed silent to me, and the house was a tall gray place, depressing looking. Still, looking in the local history book I find a photograph of her and Lee grinning happily in front of the house. Who am I to judge their lives?

Flipping pages in the same history book, I find a picture of my uncle Harold and Lee having a conversation. Harold stands with his arms folded, hat pulled down to shadow his face, looking off over Lee's shoulder. Lee faces him, leg cocked and holding a cigarette, staring under his hat brim over Harold's shoulder. Neither smiles but I know both were enjoying that laconic talk about weather or cows that characterizes ranchers to this day.

January 11. As I walk my foot crunches through the top layer of snow, then another and another until I'm lurching, ankle deep. This motion must exhaust the cattle and the wild animals.

I started writing in journals—diaries, I called them then—when I moved to the ranch, writing poems and short stories throughout grade and high school. I left the ranch for college, graduated and started my Ph.D., married a philosophy professor and singer, and created a life with him that proved unsatisfactory. After seven years, I moved home with my first husband to "repair our marriage" after his string of infidelities. At one particularly low point I moved to Spearfish, South Dakota, 75 miles away, and took a teaching job at Black Hills State College to pay my husband's child support. When I caught him cheating again, I divorced him and moved back to the ranch.

But while I was teaching, I'd met George, a divorced Air Force veteran; we were both so wary of marriage it took us nearly five years to marry. All the time, I was writing in my journals, recording my life in an effort to make sense of it. The diary entries in *Windbreak* were taken from my actual journals. Besides reflecting on my life at that time and observations of the country here, I listed the daily work I did on this small South Dakota ranch with George and my parents. I hoped, and expected, that we would live happily here for the rest of my life.

When George died from the effects of treatment for Hodgkin's disease, everything about my life changed. Though I was a widow and in my forties, my father began calling me "Child" again and telling me what to do every day. When I stopped doing what he ordered, he told me to leave the ranch and not come back. I knew his mind was not clear, but I couldn't seem to find a way to help the situation. Dozens of sons and daughters of other ranchers have told similar stories. This is one of the ways the land's vital history is lost, and why its best-trained caretakers sometimes live in exile in cities. Ranchers have not been known for keeping good records of their lives or for communicating with their families, or anyone else.

As a means of survival, I kept writing in my journals in poetry and prose. Instead of recording the details of ranch life, though, I was recording my observations about my own life. Of course, a considerable amount of what I wrote was silly, whiney, or downright wrong. I recorded feelings and beliefs that I now cannot believe once occupied my mind. But, usefully, I see patterns in my own behavior, things I later learned to stop doing.

In September 1983, my father wrote in his journal about going with my mother to Custer to play golf. He *"seemed to have trouble with the altitude"* and lay down in the car while she kept playing. He still felt ill the next day. I now think that he'd had a mini-stroke and neither of them wanted to admit it.

This began a long period during which they concealed all kinds of information from George and me. When they went to Texas that winter, my father collapsed and while hospitalized was told he'd had a stroke. Meanwhile, collecting their mail, I spotted a bill from an ambulance service. I opened it and then called them in alarm. First they berated me for opening their mail and then reluctantly admitted the diagnosis. After they returned home, Mother told me that several times during the winter my father had had "something wrong with his face." She described how all the muscles were pulled down on one side, a clear symptom of a stroke. Did they refuse to believe a doctor's diagnosis? Not go to the doctor? Forget the whole episode?

January 12. At 7:20 A.M., a chinook strikes and the temperature rises from thirty to fifty in minutes.

When I woke up this morning, I had a sudden clear vision of my father's face when he said, "I dragged your mare away." I turned away from him, crying for my beloved Rebel. He didn't tell me then that he'd asked George to shoot her or that he'd spent a couple of hot days burying her, digging a hole in the hard pasture earth with a shovel. Surely that was the strongest statement of love he could have made to me: that he kept me from finding the rotting carcass of my beloved horse. He couldn't show me that love by telling me the whole story; he could act out his love, but he couldn't talk about it. That dream is the price I pay for these days of remembering and reading journals of these past years.

I sit outside with my back to the south. My shoulders throb with the sun's heat; the downspouts begin to gurgle happily with snowmelt. As the sun melts my muscles, I wonder if either of my parents ever sat down, outside, to simply enjoy the sun, the silence. Instead of reading journals today, I will clean house, open the windows to let in fresh air, vacuum the grass and dirt the dogs track in. I will spend more time sitting in the sun. Have I learned something from all my mistakes?

January 13. The snow is gone except for shiny white patches under bushes and in a few low spots. Brown grass lies in every direction.

I remember my father telling me once that he'd managed to avoid trouble for fifty years, adding that he came back to the ranch partly so he could "slip outside and be away from it." At the time I didn't realize that might be a commentary on my mother.

Now I recall other incidents when he talked wryly about how difficult she could be. We once had a cow so wild that every time we got her in the corral for any reason, she tore down fence. Finally, one spring when she was dry—not pregnant—and had chased both my father and me when we were checking the cattle from horseback, he decided he'd had enough and went to the house to get his .22 rifle so he could shoot her.

My mother had hysterics; no, we couldn't shoot her! What a waste! Why, there were people starving in India! So on sale day he backed the pickup to the chute and we started trying to get the cow into it to take her to the auction. She knocked both of us down. She knocked down a plank fence. She bloodied her head slamming into a steel gate.

When we finally got the cow into the back of the pickup, snorting and bawling and lunging at the sides, my father said, "Someday I'm going to quit listening to that woman."

January 14. 28 degrees at 5:20 A.M. I have to wear leg warmers inside to help keep my feet warm. Jerry dresses for his shop in jeans, heavy shirt, lined vest, coveralls, a hooded coat. Then his cell phone rings and he grabs himself like a demented monkey searching for a flea, scrambling through pockets on all these pieces of clothing before he finds it.

Reading Tim Sandlin's novel *Jimi Hendrix Turns Eighty*, I was struck by two characters talking about their memories of those who have died. "How you think of her," said one, "is more important, now, than how she was." Perhaps that's what I should be doing here, instead of reading the facts of my parents' journals. Perhaps I should be solidifying my memories, especially the good ones, creating my parents in the image I need them to be. Perhaps that would let me absolve them of their mistakes, release them into oblivion.

Sleepless, I got up at 3:30 to keep reading in my father's journals, obsessed by finding evidence of his deterioration.

In July 1985 he wrote that the highway patrol stopped him *"about driving on the shoulder of the highway. Will find out what the law says."*

Neighbors had mentioned to me that he often drove to town on the highway's shoulder, because he was driving slower than the minimum posted speed. At his funeral a woman told me that one day she'd seen my parents parked on the shoulder of the road, heads leaning back on the seat rests. "I was afraid they were dead," she said. Afraid to stop and look, she called the highway patrol. My parents came home fuming that they'd been stopped by a trooper but didn't mention why, so I knew nothing of what others in the community were thinking: that they were ill and a menace on the highway.

No one has a good solution to the problem of elderly drivers. I reasoned with my parents. I talked to the highway patrol. One day not long after that fruitless and frustrating conversation my parents called me down to their house and triumphantly whipped out their driver's licenses. My father twisted his face in a sly grin and said, "I'll bet you thought we couldn't get these."

January 15. Up early again, thinking about my father's behavior in 1983, the beginning of his heart trouble and also, I realize now, the time when he began to descend into anger and unreasonable behavior, nine years before he died. At that time, George was doing nearly all the ranch work and taking medication for what was diagnosed as a thyroid problem but was in reality the cancer that killed him. This was a hard time and yet we were all working so hard we had no time to step back and think about the future.

Various studies have shown that few occupations are as physically dangerous as cattle ranching; it's been compared to professional football or work on an oil rig for its demands on the body. We were lucky; we had few serious injuries.

I am considering the future of the ranch because I am more than seventy years old, and Jerry is over sixty. When I look in the mirror, I see a face that resembles my grandmother, which pleases me because I liked her better than my mother, but still reminds me of my age.

I am considering the future of the ranch now because my father refused to make decisions about it when he could have done so reasonably and because he never included me in his plans for the future.

Recently I reviewed Susan Wittig Albert's *A Wilder Rose* about the life and writing of Laura Ingalls Wilder and her daughter Rose. One of the characters in that book observes, "Diaries and journals are dangerous because they tend to crystallize passing feelings of discouragement and bewilderments that would otherwise evaporate in the progression of events, and if they are read by outsiders, they may lead to misunderstandings." Verlyn Klinkenborg writes, "A journal always conceals vastly more than it reveals. It's a poor substitute for memory."

I must not attach too much importance to my father's journals, or my own.

I write in my journal every day. Some of what I write is of only daily importance: what to fix for lunch, what I dreamed, what we need to do the next time we go to town. Yet some of what I write might provide me with valuable insight into my own or someone else's behavior.

January 16. When I woke up this morning, before sunrise, I was startled to feel great bubbles of joy welling up in me. Maybe this is a reaction to the depression I've felt reading all these journals, all those words scribbled down recording the nuances of these past lives. Is this a sign that I am learning something worthwhile? A sign that the agony of the reading will result in joy for my life, or even for my writing life?

40 degrees at 5:20 A.M.; owls are hooting below the house. We discovered yesterday several days' worth of coyote scat right at the southwest corner of the retreat house. The owls hunt the resident rabbits from the sky; the coyotes hunt them from the ground.

Yet every night, the rabbits are active. After a fresh snow, I can look out the window in the gray light before sunrise and see them hopping around the hillside, digging into drifts at the base of alfalfa plants, or standing to nibble the dried seeds. They kick with their feet to clear snow away from the still-green flax plants beside the deck. Two days after a snow, the bunnies have beaten paths through the lilacs, from the house to the juniper trees, anywhere they habitually run. I don't understand why the coyotes or owls don't just wait for a rabbit to hop past on the bunny freeway.

January 17. My, I love reading in bed (!)—both at night, when both dogs manage to be touching me before they doze off, and in the morning, with a couple of cups of coffee and the dogs dozing again. Jerry sits in his rocking chair checking his email and playing a computer game while we sip coffee and plan our day.

Keeping up my father's weather journal while he was in Texas, I wrote on this date in 1973, *"I spent an instructional hour under the car unhooking muffler."* No details, but I can recreate the scene: the car was my 1954 Chevrolet, Elizabeth, (successor to the original 1954 Chevrolet, Beulah) and the deep snow had beaten the muffler loose so I had to remove it until I could get it properly replaced. That episode was one of many that taught me to pay attention to the noises in my car, to know what they meant.

The litany of my father's work, and the work I did while he and Mother were in Texas, could provide a handbook to raising cattle successfully on *this* place at *that* time. He did everything: kept vehicles running even though he was not naturally mechanically inclined, put up snow fence, planted trees for the future, maintained a lawn for my mother, and cultivated the garden.

Rick, who leases my land, tells me that most of the jobs we did are now outmoded by new machinery, new technology, and new methods of doing things. He says riding his four-wheelers instead of driving the pickup saves gas; he has bigger tractors and trucks, which he says are more efficient. He even sells cattle with the aid of the Internet. I tell him he's right only so long as all that gadgetry works. Should it fail, so that we had to return to an earlier time of doing more with our hands, some people who are ranching successfully now might have a hard time.

January 18. The weather has settled into January average: in the 20s at night, 30s to 40s in the daytime, sometimes with fierce winds but not much snow: good weather for the survival of both animals and humans. I can smell snow coming as moisture in the usually dry air. In summer, our walk to the ranch buildings is punctuated with sharp smells, fresh manure, sage, bruised grass. In winter the same walk is austerity for the nostrils; we smell only the dry, clean scent of grass. Unless it is about to snow.

My friend Gaydell has died, a woman and a writer I've known and admired for at least thirty years. She told her son during her last days, "I'm ready for my next adventure."

Gaydell probably began keeping journals as soon as she could write. When we traveled together, usually with Nancy Curtis after the three of us had edited one or more of the anthologies we published together, she was up at sunrise to read and to write in her journal. Her meticulous indexing and organizing of them inspired me to put mine into chronological order. Hers looked tidy and identical on the shelf; mine are several different sizes, a ragged little batch of bulging books. Whenever I visited her house, I itched to look into her journals, even though my idea of a journal is that no one touches it without the author's permission. Now that she's dead, it's not likely I'll ever see her journals or read anything she wrote in them. Even if I could read every word, they would probably not make clear to me how she achieved her self-possession, her sheer love of life. I'm sure her journal habit helped make her the woman she was. Perhaps the act of writing, rather than the words she lined up on the pages, was the important element.

So it must have been for my parents. How many people are methodical enough to write something every day? Even if a writer only records the details of a day, the daily discipline translates into something important. Precisely what each journalist learns may be different for each of us.

January 19. At dark last night the fog started moving in. When I let the dogs out at 4:30, the fog felt like soft fingers exploring my face. The fog turned pink as the sun began to rise, and then slowly flowed away east as the sun flooded the fields with light. First we could see from the foot of our hill to the corrals, perhaps a quarter mile. A dog barked in one of the subdivisions to the west, sounding so near we all expected it to lunge out of the mist. Wisps of fog slithered off between the windbreak trees. A shaft of gold illuminated grass on a hillside a mile away. Looking west, we began to see the foothills, then the trees, and finally blue sky. Magic.

January 20. Clear and cold, 10 degrees before sunrise. Every day more coyote tracks appear on the snow covering the iced-over pond south of our house: long loops crisscross from one bank to another.

Today I arranged my mother's journals in chronological order. She began writing in diaries when she was in grade school and continued through her high school years. I've found no writing done during her first and second marriages, but as soon as she married John Hasselstrom, she resumed the habit and kept it up until she died, filling cheap notebooks with quotations from poems, with leaves, photographs, and articles ripped from magazines. Conversely, she never arranged photographs in albums, just slung them into boxes; many are unidentified, which seems strange for a woman who recorded the minutiae of her days.

Learning by observing my parents writing, I started keeping my own journals when I was nine years old and expanded the habit to include a lot of record-keeping. Once I acquired a car, I kept mileage logs. Later, I added calendars, recording what I was doing in my writing or business or work life every day. With a little sleuthing, I might determine where I was on any day in the past forty years and how much gas I used to get there. All of this seems excessive and exhausting to me now; let someone else search out where I was on a particular date in 1972 if they want to; that doesn't seem a good use of my time.

As I look at the notes I took about a particular trip thirty years ago, or reread the pages of a journal where I recorded some frightening or infuriating event, though, I sometimes feel as if I am regressing into the person I was then. Sometimes I recapture beautiful memories. Other times I feel helpless anger and am lost in my own life. I can never read my journals for long without turning to a different occupation, unwilling to be reminded of the losses of the past.

Why do I torture myself by this reading? I already lived through these things once; is there any need to do it twice? Can I say that I have learned anything from living through the events or from writing them down so that I could go back and recall them thirty years later?

Yes, I think I did learn. I remember best what I have written down. From my first husband's behavior, for example, I learned particular things to beware that saved me from making the same mistake more than once— or sometimes twice.

January 21. When I think of my father, I think primarily of the lessons about ranching I learned from him while I was growing up. My brain is full of them, but I can't apply most of them because, since *Windbreak* appeared, I have lived more as a writer than a rancher, publishing a new book of poetry or prose every few years, supporting myself primarily by teaching.

Even if I got no response at all from the books I write, I would feel that getting the words in print and sending them out into the world creates a specific, physical connection between my crowded little study and the rest of the world. When I get a letter about one of my books, I'm stunned and overjoyed. My words reached someone I can name! No matter what the response is, it's heartening.

Yesterday I attended a funeral that became one of those wonderful occasions that happens at funerals: everyone visiting at the VFW club afterward, laughing and sharing stories and catching up on the news.

During the service, the minister talked at length about this farm wife and mother. Then he did something I've never seen at a funeral, but which I now believe should always be done: he introduced the woman's six children, along with their children and grandchildren. That makes such good sense; everyone there knows part of the family, but not always the connections.

He also talked specifically about life on the farm and how hard it was, and how everyone pitched in doing all the chores. He reminisced about his own days on a dairy farm and how difficult the work could be, putting particular emphasis on shoveling out the barn after every milking.

And then he made this memorable statement: "Not everything that goes into a cow comes out the tits."

I nearly choked trying not to laugh out loud, but later I heard others chuckling and repeating the comment. Most amazing funeral I've ever attended, and perhaps the very best, because it was so very real.

January 22. In my essays and poems, I predicted the events that I think are destroying the ranching economy and culture of this neighborhood. Perhaps I am studying my father's journals in an attempt to resurrect that community. And perhaps some people who read my books will work to stop those losses.

Scientists speak of a community of plants and animals that inhabit any particular environment. I believe community includes plants, animals, and people, and the land and weather. What else? Perhaps the ethics of the place, and the beliefs of the citizens.

Ethics does, I believe, have a role in my definition of community, and locally at least, I believe I see an ethical community. Many of us would not agree on political action on a broad scale, but if we see a fire on a neighbor's land, we fight it even if he doesn't vote as we do or attend the same church. If we see a car wreck, we stop to help without reading the political bumper stickers. Speaking for myself, I lie down in the American Legion hall to give blood, nodding at the woman in the neighboring chair with respect even if I clicked "hide posts from this person" on Facebook last night.

On the wall in the American Legion hall is a framed photo of Sergeant Carl Hasselstrom, the uncle I never knew. He was 36 years old when he joined the Air Corps in 1942; my father said he wouldn't have had to go, but he was embarrassed that so few from this community had volunteered. Carl was a member of the 99th Bomb Squadron. On his second mission, in April 1943, his B-17 bomber was hit. He was the only crewmember killed and one of the first seven casualties buried in the National Cemetery in Sturgis. Ethics.

January 23. I woke this morning hearing a cow bawl nearby and thinking I needed to get up and start feeding; took me several minutes to realize I no longer own cows. I remember vividly how we spent most of every day caring for them; my body misses the work and I feel guilty not doing it.

In my journal for this day in 1975 I wrote of a conversation with my uncle Harold Hasselstrom. He was having midmorning coffee, sitting at the counter eating Aunt Jo's chocolate chip cookies, when I walked into their kitchen. As usual, he wore a blue denim shirt over a white t-shirt, with striped suspenders. His graying hair was smoothly combed to one side and creased from the broad-brimmed hat hanging by the door.

We talked about the snowfall and how the cattle looked. I was worried that the roads might be icy that night when I was supposed to attend an environmental gathering. Harold agreed; he had a school board meeting.

I ranted; why, at age 72, did he have to waste his time with those notoriously argumentative locals?

"I sympathize with you for spending forty years on the school board," I grumbled, "and you don't even have kids."

He raised his coffee mug and looked at me over the rim, one eyebrow raised, blue eyes twinkling.

"Somebody has to do these things," he said. "I haven't wasted a minute."

That's why there is a gymnasium in the Hermosa School named for him, because he selflessly donated his time, energy, and money to the children of this neighborhood. Ethics.

January 24. I nearly sent an email to Gaydell today, with a play on words I knew would make her laugh. We didn't email often, but my experience tells me at least a year will pass until I remember, before automatically typing her name, that she is dead.

Emails: the disappearing journals of the present. Like all journals, they are filled with mindless blather, but they also contain wisdom, pathos, empathy. No one will be able to dig through them to separate the pearls from the stones. Well, maybe, if the government or someone else is eavesdropping on all of our communications.

January 25. A low of 21 and a high of 48 but the wind howled so loudly I kept thinking we must be having a blizzard, typical for this time of year. Instead, it's a chinook.

When journals are published, whether they are of private individuals or public figures, part of their value is often how they comment upon events that occupy the state or nation at the time. My father's journals rarely mention news. *"John Wayne died"* he noted on June 21, 1979, and on July 29, 1981, *"I am tired of the British wedding."*

Yet in 1984 he wrote, *"We are happy with the election,"* adding a few days later, *"I should mention that Reagan won by a large margin and Linda's group's nuclear dump initiative won handily."* With that entry, he proved to me that he was acknowledging his awareness of an audience. Did he think that he might read his own journals?

Or was he thinking of me, even though he later called my writing "junk?" In order to explain that later behavior, I must believe that he lost his mind, and because he had not prepared a will that allowed for that eventuality, he acted against his own best interests.

Usually his journal comments were more local: *"country starving for a drink"* he wrote, when we'd had no rain for months. Later, *"This hot dry weather is getting to be an endurance contest."* This was as close as he came to complaining.

January 26. I am immersed in both sets of journals. My father apparently began keeping journals not long after he adopted me and married my mother when he was 44 years old. My biological father was Robert Paul Bovard, but by the time I was four years old, my mother had returned to South Dakota, divorced. During the summers, when school was out, I lived with my grandmother in Red Canyon while my mother worked in Rapid City until she married John Hasselstrom when I was nine.

After George died, and my father began treating me like a child, I moved to Cheyenne, Wyoming, where I lived for almost seventeen years. In 2008, after the deaths of both my parents, I returned to the ranch with Jerry, who had just retired from the Wyoming Department of Transportation.

January 27. Reading my father's journals has flooded me with good memories, but they have also shown me some of the hardships I missed because I was away at college or married. I'm beginning to see some of the events that resulted in his gradual change from a loving father to an angry stranger. During his last twenty years, he often said, "Life isn't supposed to be fun." Yet he decided that he needed to "learn to play," as he put it. So he bought some used golf clubs, and he and Mother went to various courses to hit balls. He also carried a club when they walked in the fields in the evenings and killed several rattlesnakes with it. I still remember the sheepish look on his face when he held out a mangled snake to me on his golf club, saying, "I thought you'd want the rattles."

I can't read my father's journals for very long before I become overwhelmed by anger or sadness or frustration and must find other work to do. Sometimes I sense some great lack in what I know or I'm swamped by the pain of recalling those days. On such days, I try out a new recipe for lunch, or bake bread, my brain as busy as my hands as I knead the dough.

Why should this reading depress me? Because I cannot correct any mistakes now. Whatever he thought, I can't change it. So I must either let go of any pain it causes me or find another explanation.

January 28. At 6:50 this morning, the sky is completely overcast. The air feels muffled, as if a cloud full of snow is hanging overhead. An owl hoots from the retreat house and another answers from down the valley to the east.

Tonight, both of us exhausted from our respective days of work, we had popcorn for supper. My mother wrote that making popcorn was "Linda's favorite indoor sport," but I enjoy it even more now. Mother was stingy with butter, preferring margarine because she thought it was healthier; I hated the greasy stuff. Jerry serves popcorn not only with generous amounts of butter but plenty of Parmesan cheese. I keep a bottle of hot sauce handy to liven it up after I toss a few kernels to the dogs.

January 29. Without a snow cover, the land looks particularly brown and dusty. The pronghorn match it perfectly: we can look out the bedroom window and declare, "No antelope this morning!" Like most people on the plains, we use the terms "antelope" and "pronghorn" interchangeably. Five minutes later with a shift of the light, we see six of them grazing, their white bellies and tawny coats blending in with the dry grass.

"The antelope have materialized," we tell each other.

As we step out onto the deck, we hear a sharp sound, like the *yap!* of a small dog as one antelope barks to warn the others. All the heads snap up and they watch us. Even if we quietly go back inside they may take fright and dash off.

January 30. At three this morning, coyotes were howling so close to the house that both the dogs woke, pricked up their ears, and stared at the windows behind the bed, eyes wide and shining. Now, at nearly sunrise, the sky is covered with clouds with only one clear strip on the eastern horizon, so as the sun comes up, it reflects off the cloud bottoms and turns the whole gray sky a glorious golden pink.

Peeling potatoes for lunch, I was thinking of possessions and how they can carry memory when I realized my potato peeler is almost as thin as the one I took from my grandmother's kitchen after her death, worn nearly to wire.

Patting a chicken dry before roasting it, I realized my hands are spotted with age, just as hers were the last time I saw her do the same thing. My mouth waters, remembering the flavor of the chicken she fried in lard and the gravy thick with cream and chunks of fried chicken skin she spooned over mashed potatoes to go with it. She'd butchered the chicken from her own flock. Mine comes from a neighbor but as I slice onions I can catalogue my grandmother's utensils still in my kitchen: her old potato masher, a couple of butter paddles, a wood-handled can opener that feels warm, as if her hand had left it only a moment before. I gave her bread bowl to my cousin Sue; I hope her son will someday cherish it.

These ordinary objects are the record of how she lived her life and her legacy, a treasure so worn no second-hand shop would take them if I discarded them.

January 31. When we go without rain or snow for a couple of weeks, we whine about drought, but these grasslands were formed by prolonged episodes of such weather. Grasses can wait a long time for rain; deeply-rooted grass can survive long dry spells by drawing water from underground. My father said it takes range grass three to five years to recover from drought or overstocking, so even if we get good moisture this spring, ranchers will need to limit the number of cattle on the pastures by selling stock they'd prefer to keep. The drought is doing all the talking now.

February

February 1. In 1975, I wrote in my journal, *"I have a new feeling of unification, consistency, on the ranch, as if it doesn't matter if I refinish a chair, cut wood, fix fence, work on my writing, because it's all work involved with my unique place in life and spot on earth."*

This is exactly how I feel now—but why did it take so long for me to rediscover this truth I learned nearly forty years ago?

Those who believe in reincarnation say we live the same life again until we learn how to live correctly.

Is this the structure of life, to keep rediscovering the same things again and again? I've often joked about finding by reading my journal entries that I'd made the same life-changing decisions over and over, but this is a shock. Perhaps this scrutiny of my journals and those of my ancestors will provide me with an answer at last.

February 2. I have a new theory about the folks who drive by here on the highway. They believe that they can eat any snacks without gaining weight *if they eat in the car and throw the package out the window.* Today while walking the dogs down to the ranch buildings, a half-mile from the highway, I collected papers that once held dill pickle-flavored sunflower seeds, sugared mini-donuts, and a hamburger and fries from a place decorated with golden archways. A day rarely passes when I don't come back from any walk with my pockets stuffed with such trash.

This leads to my second conclusion about drivers: they see nothing beyond their windshields or the highway. Their destinations are the "scenic wonders" of the Badlands or Yellowstone, but until they get there, they are blind. They don't even realize that if our economy was correctly conducted, that hamburger would have come from cattle eating my grass instead of from a scrawny Mexican steer fattened on hormones and herbicide-laced corn while standing knee-deep in its own wastes.

February 3. Snowing at 5 A.M., 15 degrees. My dad always said "Little snow big snow," meaning that snow falling in small flakes would be deep. Rolling clouds of snow follow the big trucks passing on the highway.

The grass glows vivid gold because the snow's moisture has soaked into it, softening it, making it more palatable for the cattle. They graze happily on the gleaming grass.

Unlike many humans, cows know how to use their resources as they find them. These cattle live much as I imagine wild cattle did, free to roam through a huge pasture because Rick has torn out a lot of fences. During the winter, they stay fairly close to where they are given supplementary feed. In the summer, though, they graze near our house for a day or two and then wander off over the hill to graze a mile or more away, though they are in the same pasture. After a week or so they come back to this area and eat fresh grass that has grown up since they were here before.

At some point in every day, whether I'm in the shower or fixing lunch, I give myself a pep talk about writing to keep myself from devoting my entire day to something I consider easier but of less worth, like cleaning house or dusting. If I spend too much time on these chores, I feel guilty because I haven't written. Today I lecture myself while making the preparations for turkey tetrazzini; I tell myself that it's important to get back to writing on this journal about the prairie grassland, even if I don't finish it, even if no one ever reads it, because it is helping me to understand and accept all the disparate elements of my life.

Maybe the larger question is, why should I feel guilty about writing? Not only is this my chosen profession, but sometimes also I get paid for doing what I love, putting words together in intriguing ways. Not much and not often, but if payment is what my WASP upbringing requires to make a job legitimate, this one is. I teach and learn what I use in writing to teach more effectively.

My parents, despite their respect for reading and a college education, taught me that WORK in capital letters did not include writing or sitting down. If I was not physically active, moving, riding horseback, hoeing weeds, vacuuming, I was not really working. In spite of all my talk about the legitimacy of my work to others, in spite of believing the truth of what I say, I still have that shadow of parental expectation over me. Does everyone feel this way? Or only those raised by workaholics?

February 4. A rough-legged hawk was soaring this morning, spooking the grouse into chuckling flight before it landed in a tree and eyed The Bad Breath Boys—the dogs—as we walked by.

Leaving the post office, I wave at a woman I barely know who is parking to get her mail. On the sharp curve as I leave town, I wave at a man I don't know, but who gives me a nod and a two-finger wave in return. I wave every chance I get, because I like living in a place where neighbors wave at each other; if the newer residents don't know they should wave, I hope I am teaching them.

February 5. I have been fighting a cold for several days so I got up about three this morning to sit in my reclining reading chair to read and cough more effectively.

My creative thought takes different forms, depending on the necessities of the day. I am alert to my surroundings, noticing everything in case it might become a part of what I am thinking or writing. At least that's what I told myself today when I noticed it was time to clean the toilets and spray my homemade grease-busting concoction on Jerry's shop overalls before I laundered them. As I sprayed, I remembered we are going to town tomorrow, so I began to make a list of the jobs we need to do: deposit checks in the bank and credit union, return books to the library and check out more, buy lumber for Jerry's latest project, oh, and take the recycling.

Eventually, I get back to the computer and realize that while one level of my mind has been occupied with these mundane tasks, I have recalled an idea I want to explore in a poem, so I make notes.

February 6. This morning at the retreat house I noticed part of one of the green plastic gutters lying in the yard. I picked it up to put it near the remaining piece so that on a warmer day I can put them back together. How, I wondered, did it break?

Then I realized that the end of the plastic gutter still attached had big teeth marks in it. Looking inside, I could see what I thought was a branch with shredded bark. So I pulled on it.

Not a branch. A rabbit foot, bones exposed and gnawed.

The rabbit had evidently run into the gutter to escape the coyote and become wedged, trapped while the coyote gnawed on his foot. I tipped the piece over and looked into the rabbit's frozen furry face. Perhaps the rabbit died of fright or shock and pain, or froze to death.

February 7. Toby jumped a rabbit in the ranch yard and went yipping after it, followed by another rabbit, apparently confused about how these things work. Cosmo swung in behind the second rabbit, and they all disappeared into the corral in a flurry of snow and dust. Needless to say, when the dogs came panting back, they weren't carrying rabbits in their jaws.

Leaving the post office today, I drove slowly down the street behind three children and their mother on bicycles.

She didn't look behind her until we'd gone four or five blocks, then suddenly glanced at me, shouted to the kids and veered to the shoulder of the road shouting "Sorry!"

I rolled down my window and said, "Don't apologize! This is how small towns are supposed to be!"

We've had high winds all day, and the snow is so nearly gone that the fire danger is terrific. Gusts are 55 miles an hour now, and will reach 70 tonight.

Today I looked at my mother's diary for February 1957. She wrote enthusiastically about taking me and a girlfriend to a 4-H dance and rhapsodized about full skirts, lacy blouses, and voluminous petticoats that the dancers wore.

"Just this morning," she added, *"Linda told John she was already too old (at 14) to learn to dance but she seemed to catch on pretty well—ha!"* Poor woman; she was so pleased when I seemed to share the things she cared about. I tried, when I was young, to be excited about dresses, but I couldn't do it very well. She always bought me pink clothes; I hated pink and still do.

When I got married the first time, she insisted, at the top of her lungs, that I choose a silver pattern, registered it at a particular store and promised to buy whatever pieces to an eight-piece set I didn't get as wedding gifts.

I got two spoons; she never bought any more. Now I have the silver service she acquired in one of her marriages. Before special occasions, she'd put me at the dining room table with a soft rag and silver polish. I use the set every day. When tarnish appears, I heat the whole set in a pan with baking soda and crumbled tinfoil. Every day we eat enjoying the weight of the utensils. Mother would be horrified. But why save it for special dinners when we rarely entertain? And who am I to leave it to?

Walking the dogs, I wandered into the greenhouse, showing 80 degrees while it's 60 degrees outside; last year on this date we set a record high of 75. Last fall, as plants withered and died, I brought the pots inside the greenhouse, so shriveled Thai peppers, marigolds, lavender, and oregano wait to be discarded. In the rafters are tomato cages and stakes; empty pots crowd the shelves. Above my head is a pair of gray and brown grouse wings discarded in the windbreak by a coyote or owl. My gardening bucket stands on a shelf, holding all the tools that get so much use in the summer. And from one pocket I pull a pencil and paper to make a note about the poem I worked on this morning.

February 8. S—, who is here on retreat, laughed when we both pulled tissues from our pockets for our runny noses. "Remember," she said, "how little old ladies always had pockets full of tissues?"

Instead, what I remembered in that instant was that my grandmother always had embroidered cloth handkerchiefs, now in a trunk at the foot of my bed. How she must have loved paper tissues when they became available, so she didn't have to wash and iron those handkerchiefs. When I was horrified that she had thrown away some of her kerosene lamps and rescued some to take to my own house, she said, "I lived with that old stuff too long; I don't want it anymore."

S— and I talked about how we are becoming the older people we stared at and failed to understand and couldn't imagine becoming. We have more gadgets, more access to news, but are these differences significant? Do we simply turn into our ancestors as we age, having learned nothing in a generation? Surely not.

After I left S—, I continued to think about age. What do we assume about age, and thus take into ourselves? What influences our behavior because it's what we're told or shown? When we lived in Cheyenne, I noticed the wives of Jerry's colleagues, mostly in their thirties and forties, begin to say things like, "Oh I'm so old!" or "I'm so fat!" I always argued that such talk would encourage them to become what they feared. But TV, social media, magazines, and movies encourage such thinking. Public attention is focused on the young, lithe, athletic. Our heroes are sports figures, movie and TV stars, and some class of people called "celebrities" who apparently do nothing but spend money to look better.

Even news stories sometimes seem slanted against the elderly. Nearly every story of a traffic accident involving two drivers of differing ages begins with the age of the older driver: "A car driven by an 85-year-old man collided today. . . ."

If my comments are not merely whining, they must lead somewhere. My choice is to speak out against these assumptions every chance I get. As a Tim Sandlin character says about age:

> "What makes us unique from other minorities, such as blacks or homosexuals, is that we were once like you and, someday, if you survive, you will join us. A landlord or employer would never treat a black man with contempt if he was certain of becoming one. How can you treat the aged as nonentities, knowing where you yourself are going?"

February 9. Tonight a huge fire is burning west of us in the Black Hills; the smoke rolls black against the sunset, impenetrable. The folks in the subdivisions west of us, close to the trees, should be feeling nervous but perhaps they can't see the smoke. Or perhaps, having lived in cities, they confidently expect to be protected from any danger. They may not realize that much of the firefighting here relies on volunteers, who are mostly from the surrounding ranching families. They are dedicated, but many of them haven't much experience in timber fires. No doubt the forest officials are calling in experts, but there has been considerable criticism lately of the way the professionals take time to get organized before they actually start attacking a fire. The locals dive right in, risking everything for their neighbors.

Throughout the evening, I keep thinking about those subdivision residents, wondering how many of them know about the fire or have any sense of their own danger. Pioneers who moved into this country built their own homes and that took time. Most homesteads took several years of hard work to develop, but settlers learned about the land, discovering where they could shelter from windblown snow, where the best grass grew and therefore where to put a garden. In a few years, they learned their piece of land and how to live on it best, so they adapted to their surroundings.

Today, folks who have always lived in cities can buy land, choose a house site for the view, and hire builders so they can move in without knowing anything about their land.

A geologist friend who was conducting a study for a college class in the nearest subdivision said residents came out of the houses to ask him where their land ended. But, as Ann Haymond Zwinger remarks, "Ignorance is no defense in the court of natural law."

We speak of ownership of land so casually. The new holders of the subdivision houses received records from a title company; they wrote a check or signed a promissory note and got pieces of paper that declared their ownership. Probably they didn't give much thought to the original settlers, whose signs on the land are small: a few dusty trails, barbed wire fences, cattle tanks, leaning fence posts. The Indians who preceded them left few signs they'd recognize. Neither did the pronghorn, deer, porcupines and coyotes who don't even know that their title has been terminated.

Several of the new residents have put up "ego gates," those huge arrangements of uprights and crosspieces few ranchers ever erected. Worse, they've put up "security" lights that glow a ghastly orange all night, driving away the animals who formerly hunted in darkness on these ridges and valleys. Some of those folks moved to the country saying they wanted to be near "wildlife." Their neighbors wonder aloud why they are so afraid of the dark.

Still, a rancher friend says, "If those houses were empty, they'd fall down in a few years, just like the homestead places did."

Perhaps one of my most important reasons for writing is to record what this country looked like before newcomers destroyed the darkness. Will I ever again see the real Northern Lights, those long ribbons of blue and green and red flowing and flaring like great sails in the sky? We can light up the dark if we need to, but ours operate only when we need them, with switches that stay off most of the time.

February 10. The sun rises through a murky yellow haze that makes the grass look yellow, all from the smoke of the burning forest fire. My throat is raw and the bitter flavor of scorched pine is on my tongue as I wake.

When I turn on the light over the bed, Toby untangles himself from the blankets, snorts, sneezes, and looks up at me, yawning. He stretches himself out as long as he can while I scratch his ears. Meanwhile, Cosmo has hopped from the foot of the bed to the floor and is scrubbing his face on the rug, snorting, and waking himself up. Instead of springing, I step carefully out of bed and do my own slow stretches.

The horned larks are back! Their songs twinkle in the sunrise. These are the birds I called "the invisible birds" in my book, *Windbreak,* because I didn't know what they were. By the time I wrote *Between Grass and Sky,* I'd learned that they are horned larks. The males wear nifty black masks that extend back into feathery "horns," hence the name. About the time I learned to identify them, we had a series of wet years and they vanished because they prefer dry shortgrass plains. I mourned their loss, even as I was grateful for the moisture that drove them away.

Now that the drought has broken deep cracks several inches wide in the dry earth around the house, I am seeing great flocks of horned larks flying from cedar trees to the dam to the prairie to the cottonwoods, all the while twittering in tiny voices even on the coldest days.

Walking the dogs, I gasp as I step into a pool of cold air. At the same instant, I hear babbling and look overhead at geese going north. "Too soon!" I shout, but they pay no attention.

February 11. At sunrise I realized the Black Hills were covered by a cloudbank so blue and thick it looked like sky. Then the snow began, fine as pinheads. When the sun shone through the clouds, it was accompanied by sundogs, glowing spots on both sides of the sun. Some say the phenomenon got its name because the shining spots are as close to the sun as a dog to the heels of its master. Ironically, while I was looking up, I tripped over Toby.

I draw my bedroom shades at night to close out the glare of the subdivision lights. Each morning, when I raise the shades I see a line of cars leaving those houses to race up the four-lane highway built mostly for their convenience, going to work. After dark they race back again, having spent all day in the town they left to live in the country. If we drive past their houses in daylight, even on weekends, we rarely see anyone outside.

Their lives are not my idea of living in the country, nor are rows of houses the highest and best use of this land, adapted over countless millennia for grazing animals. I've read that more than half the world's people are urban, which means humans' understanding of the requirements and benefits of rural life may continue to decline.

February 12. Beside the cattle guard we found an owl pellet, mouse fur wrapped around a miniature jaw. The owl must have regurgitated the leftovers of last night's meal while perched on the ornate gate Jerry made.

The divided four-lane highway is one of the most obvious changes in the ranch, but not the only one. I opposed it by suing the state, a battle I lost, naturally. Before the new highway, we had an underpass so we could drive cattle from our pasture east of the highway to pastures on the west side. Now the only way to move cattle to that pasture is by truck, which severely limits the use of the grass. Jerry's career was with the Wyoming Department of Transportation, so he knows if this highway had been built in Wyoming, the state would have been required to provide me with an underpass. South Dakota politics often operates against citizen interest, so the state reduced the pasture's value to me without compensation. With subdivisions on all sides, the county keeps trying to rezone my pasture as development land. I will continue to resist, though I've also threatened to build *BeLinda's Best Bar and Brothel* over there.

February 13. Every night I look up at the stars and trace pathways. I don't know the names given to constellations by those who have looked at them for thousands of years. When some kind soul has pointed them out to me, I simply can't see what everyone has agreed is there. Orion and his belt? Pegasus? I can see the Big and Little Dippers and pick out Venus when it's bright, but that's about it.

I don't understand why people want to see familiar shapes in the sky patterns. These isolated stars have no desire to be part of a picture imposed by puny beings on this planet. They are alone in their glory and don't need our names or permission to shine. I prefer to create my own pictures. Leaning against the deck railing, I hope for rain to spill out of the dipper. I picture

George striding along a glowing pathway beyond the bright stars, his eyes on a horizon far from this one. Perhaps that flickering glow is a message we cannot read yet. That great oval of darkness: what is out there, so far that we can't see it?

February 14. Valentine's Day. 10 degrees with a cold wind at 5:26 A.M.

VALENTINE FOR MY MOTHER

I.
Cut flowers don't last
says a woman's voice.
I spin around in the Safeway aisle
expecting to see my mother
who's been dead all winter.

Cut flowers don't last,
she says again,
the woman with blue hair
beside the flower display,
shaking her head at the young man
still reaching for a bouquet
wrapped in red paper.

She sounds like my mother,
mouth pursed, not smiling,
each time I brought a bouquet
to the nursing home. You shouldn't
have spent the money, she'd say.
Cut flowers don't last.

I picked them
from my garden, I'd say.
She'd snort.
Cut flowers don't last.
So I brought slips
from my plants,
potted them for

her window sill. She didn't
give them water.

II.

When I was growing up
Mother served our meals on Melmac
scrawled with scratches,
kept the good china
in the cupboard
so it would last.

During that final year
she was alive, she asked once
about her good china. Safe
in my glass-front hutch, I told her.

At ninety-two she took her final breath.
I covered her pink enamel coffin
with roses the color of every blouse
she gave me no matter how many times
I told her I hated pink.
As I paid the florist
with her money, I told him
Cut flowers don't last.

III.

Now in the Safeway aisle
I smile at the young man
who is carrying the flowers
toward the checkout stand.
Cut flowers don't last
she says once more.

Tomorrow all the blooms
that do not sell will pucker
in the dumpster
brown as the roses whipped
by the cemetery wind

the day after my mother's burial.
Cut flowers don't last
I muttered to the mound
above her heart.

IV.
I gave her dishes to my cousin's
daughter. In my gardens,
I cut flowers, thinking of my mother.
Blooms scent every room,
reflect themselves even
in the bathroom mirror.
Every night from the arbor
I watch the sunset
that will never come again.

I've worked on that poem a long time, half embarrassed because of its negative mood, but it expresses feelings I've carried for a long time too, and my recovery from them.

February 15. A great day. I worked on an essay about the prairie sky, conducted a Writing Conversation by email, and made pork fried rice for lunch.

All day, I consciously resisted looking at the journals piled on my worktables and beside my chair. And all day, whenever I looked down at the ranch buildings, I thought I saw my father just stepping into the corral or my mother shaking a rug on the porch.

Even though I am not riding the range every day, I am in and of this ranch. I move through my days accompanied by thoughts of those who worked the land here before me, the people who are responsible for my being here. They will all be part of whatever decision I make about the future of this place.

At night, I avoid looking north, where the bold lights of Rapid City shine upward, blotting out the Milky Way. If I stand on the south side of the house I can still look south and east down the L7 draw into starlit darkness, where the coyotes, owls, and probably mountain lions can still hunt. In daylight, a heron stalks Lake Linda, gobbling frogs and occasionally killing a duck. Ducks and geese graze along the dam's edges. Cattle graze the hills and lie in silhouette on the top; they are no longer mine, but they belong to a ranching family and they truly express the nature of this place. I can take satisfaction in that.

Not all the changes in the neighborhood are destructive, I must grudgingly admit. On the west, between my house and the ugly tin sheds of the newcomers stands an ancient cedar log cabin, headquarters of the Great Plains Native Plant Society. For a decade I have worked with its members to establish a botanic garden dedicated to shortgrass native vegetation on 350 acres of land I loan to the group. Now we are working toward a plan to help the GPNPS own the garden, in memory of botanist Claude Barr. Perhaps some of this prairie I have loved will be available for generations beyond mine, even though making it so requires me to cooperate with creation of a parking lot, for example. Inside the tiny building, two large EXIT signs shine all night long, evidence of regulations required for public access. Compromise; that's the answer I have found for the present.

Dan O'Brien, a wildlife biologist, buffalo rancher, author, and my longtime friend, says that the grassland of our area, Custer County, is one of the greatest unrecognized treasures in the nation. In open pastures surrounding my ranch, two privately-owned buffalo herds browse grasses that are the best example Americans have left of plains habitat as it naturally evolved. Feeding cattle grain in feedlots is destructive of plains habitat because, to grow grains, farmers may plow up native prairie on which bison thrived, along with a rich population of other flora and fauna. O'Brien has observed many plains landscapes where plowing has created a farmed monoculture. His ranch, and the nearby 777 Ranch, along with the GPNPS garden, might preserve a glimpse of that incredible resource and perhaps prevent us from losing more of it.

February 16. At 5:10 A.M. it's zero and breezy. Reading Barbara Kingsolver's *Animal, Vegetable, Miracle* in bed. She suggests that we begin teaching in school classrooms how food is produced, arguing that the story of how bread begins in tilled ground and arrives at our tables is as important to our knowledge as the history of the thirteen colonies. "Couldn't one make a case, she asks, "for the relevance of a subject that informs choices we make daily?" She notes that ignorance about our food sources causes additional trouble: overdependence on petroleum and diet-related diseases like obesity.

I agree, yet the average American probably knows considerably more about the origin of bread than how meat is produced. Evidence of this can be found daily in social media statements about "rich ranchers" and how they "control" the West.

Walking the dogs through this morning's heavy snow, I found a series of tracks that told a story I believe I've interpreted.

A rabbit was hopping toward the oak tree when an owl or eagle struck, leaving a blood spot and marks of wings. The bird walked several steps, dragging the rabbit and leaving a mark in the snow that looked like a feather fan—its tail dragging. The tracks it left were five inches long. Then came a large blood-smeared spot where the rabbit was partly eaten, and one final set of tracks where the predator bird took off.

We've seen both a great horned owl and a bald eagle within the past few days, but the tracks looked like the photographs I saw when I searched for "bald eagle tracks in snow."

February 17. 25 degrees at 5:26 A.M. when the dogs leap to the floor.

I just found my report cards from grade school, when I was Linda Bovard. My kindergarten teacher classed my application and deportment "very good" and called me "an excellent little student." In first grade I received an S for "satisfactory" in everything: writing legibly, writing with reasonable speed, enjoying good music, muscular communication, reading with understanding. During the school year I was never absent and never tardy, a tribute to my working mother's organization. I got a C in reading during the first six-week period. After that, my reading grades were all As, with Cs in penmanship.

By the time I was in fourth grade, Mother had remarried, John had adopted me, and I attended Hermosa School as Linda Hasselstrom. The lower room teacher, Elsie Enders, recorded my year's average for arithmetic as a C+. I continued my art ineptitude with straight Cs, but got an A in reading and a B in spelling. A "good little girl," I always did well in the subject called "Deportment"; are students judged on that anymore?

What a privileged child I was. One of my parents always signed my report card, but first they both questioned me about what I was learning, talked with me every night about my schoolwork, and insisted that my jobs were both my homework and my chores. They took seriously their roles of raising me to be a decent human being. When I read about child rearing today, I realize again how fortunate I was.

February 18. At 5:23, the temperature is 25 degrees with snow; the radio says visibility is especially bad on Highway 79. Our pantry and freezer are full, so we have no need to go anywhere.

According to the most recent census, nearly 81 percent of the U.S. population lives in urban areas, with only a little more than 19 percent living rurally. Of course this trend has continued throughout my lifetime; in 1940, the figures were about 56 percent urban as opposed to about 43 percent rural. In 1950, when I was in grade school, the statistics were about 64 percent urban to 36 percent rural. What this means economically and politically, of course, has been well documented: fewer workers depend upon rural businesses like ranching, farming, and their related occupations for a living. Fewer people arrange their day's work depending on the weather; fewer Americans produce the food we all eat.

Understanding what this means for me as a rancher and writer has taken me a long time; I can't name the time when I began to realize how our population was becoming urban. When I attended high school in the late 1950s, I saw contempt for the rural kids from those who lived in town. When I missed school because our ranch road was closed by snow, the teachers in the high school 25 miles away weren't sure my excuse was real. Other changes went deeper. I began to see more articles saying that fewer people were eating beef because they believed cattle grew up or were fattened on hormone-laden grain in stinking feedlots owned by multinational companies.

One of my former students recently persuaded his book club to read my *Going Over East: Reflections of a Woman Rancher,* my second published nonfiction book, which originally appeared in 1987.

I was delighted that a book group was going to discuss the book and eagerly anticipated their reaction to what I had written about how we managed our ranch to preserve grassland, with details drawn directly from my journal.

They hated it.

Why?

They agreed that it couldn't possibly be true. They thought life couldn't be that hard and still be endured; surely I had made up the experiences in the book to make it appear that I had a difficult childhood. Another reason I was clearly lying, they said, was because a person would never choose to go back to a life so difficult, as I did when I returned in an attempt to repair my first marriage, and again when George and I moved back to the ranch. Their final argument was that if my father was a committed rancher, he would never have taken time off to go to Texas in the winter during the final years of his life.

In writing to tell me about the meeting, my former student added, "So be aware, Linda, you are a lone voice crying in the wilderness of South Dakota. . . . People don't want biography in the form of gritty autobiography. They want literary pornography instead; something they can't smell."

My friend and correspondent tells me that several weeks after the book club discussion, one of the book club women accosted him in the grocery store and told him that clearly he did not understand what the club wanted to read and he should not have suggested such a book. He said that he had worked as hard on a farm when he was a child, feeding hay to horses and cutting wood and doing other chores.

"I don't believe you," said the woman, "because that would have been child abuse."

At first I was stunned and angry, but thoughtless anger is precisely what's wrong with a lot of the discussions occurring in America today. I prefer that people who disagree do so with civility, and with reason.

February 19. 10 degrees at 4 A.M. with colder weather predicted.

Tonight, I stepped out on the deck after my bath to cool down and heard the mellow hoot of a great horned owl from the trees around the retreat house. Above me, the handle of the Big Dipper pointed straight down, which, I told Jerry, means we'll get moisture. A realist, he snorted; "It's winter; of course it points down."

Thinking this morning about the readers who didn't believe *Going Over East.* When I wrote that book, in the 1980s, most of my neighbors did the same jobs the same way, and many of them worked harder or suffered more. My grandmother regularly killed rattlesnakes and skunks with her garden hoe. Some of the great-grandmothers in this neighborhood would have considered my life a vacation. These tough people usually had neither the time nor the inclination to write about their lives; my life was a pale imitation of what the pioneers endured.

And what about the well-documented hardships of working people during the Dirty Thirties?

This book group has my sympathy and my pity. Maybe they've watched a lot of "reality" TV. My newspaper regularly informs me about lying politicians, business leaders, and ministers. Perhaps the book group members are so used to a daily diet of lies they no longer recognize truth.

A responsible reader might try to discover evidence that I am a liar before announcing that I am. In recent years, ranching has been the subject of many written and televised features, so research on the reality of what I wrote wouldn't have been difficult. I had hoped they would respond to me with questions so as to expand their knowledge of cattle ranching and grasslands. I find it alarming that they could so easily dismiss the truth about the hard work that is part of producing the food we eat. If they are so immersed in media lies as to dismiss the book's simple truths, they will be more easily hoodwinked by companies who want to control food production in this country. That, not my personal disappointment, is the primary danger of their disbelief. I can't predict every response, so I will keep writing what I know to be true.

February 20. I've been reading about "ground-truthing," a term used in meteorology and some military applications to describe what I think of as "reality." First, aerial photography, satellites, or other sources provided views of a particular location. This view is then checked by ground-truthing: people are sent to that place to measure, observe, and take samples, and their information is used to verify what is shown on the high-tech images and to update maps. Without the verification supplied by people actually visiting a particular spot, the high tech data is hypothetical, at best.

Surely this conclusion is self-evident and applies to other disciplines as well. Over and over in politics, we have urged the folks elected to govern us to come home and see what's really going on, instead of staying in the state or national capitols talking only to each other. Sometimes it seems as though they rely only on "data" that appears to have been collected from outer space.

Moreover, that's what I'm doing living here: finding my truths on the ground, in the land where I walk every day, where the food I eat emerges from the ground.

There's a deeper implication here. If fewer of the world's people are living in rural areas every year and the majority live in cities, then the truth of rural ground is being lost. Daily, our society gulps down a diet of images, tweets, rants, observations, and pronouncements that have not been tested against reality and that have nothing to do with rural life. The average person may not walk on any dirt at all during the day: she might go from wood floors to concrete sidewalks to automobile carpet to sidewalk. Even if she cuts across the office building lawn, she's not really in any meaningful connection with the ground.

How can we intelligently manage rural areas, including wildlife habitat, parks, and other protected areas, if we have no idea of the truths known by those who walk over that ground? Yet every day politicians and business people protecting their own profits overrule the scientists who are nose-to-nose with the grizzly bear of reality.

February 21. 15 degrees at 5, light snow, a little wind. Rick is feeding cattle now, rolling out bales of hay weighing several hundred pounds every other day. The day they are fed the cows stay on the feed ground, eating and napping. The next day they get out and graze, ranging widely into several pastures. The third day the cattle drift back toward the field where he will be feeding again, not on top of the old hay but far enough away so the manure is scattered.

February 22. One July day in 1954, at age eleven, I wrote, *"Kittens marooned, but I rescued. Darn near drowned doing it though."*

I say that I write nonfiction. Since I've begun publishing my work, I've struggled to be as honest as I can. I've recalled conversations between myself, my husband George, and my father, repeating what each of us had said, over and over, trying to be accurate. But memory can lie, deceive, and even subtly shade the scene so that the writer appears to be wiser and nicer than she really is.

So the truth is that on that day in 1954, a terrific rain and hailstorm flooded the alfalfa fields between our house and the railroad tracks. The kittens had scrambled onto a fallen tree a quarter mile from the house, where we heard them meowing piteously as we were inspecting the damage. I rode my horse to the tree through knee-deep water, stepped onto the log, stuffed the kittens into my jacket, and rode back. The kittens meowed and struggled and scratched. The horse, flicking her ears back at the racket, pranced and splashed water. But my worst danger was of being scratched or thrown, not drowned. And my parents were watching.

Perhaps in adamantly declaring I tell the truth, I'm still compensating for the exaggerations of childhood. Who did I expect would read that entry and believe it? Was I writing for some posterity I expected to hoodwink? I feel myself blushing.

February 23. Wild weather today: over 40 degrees in the morning and we saw a V of geese going north with one goose far behind the others, honking and flapping. This afternoon gray frothy clouds blew in from the west and north, tossing the cedar branches.

Jerry was tucked into his shop all day while I was busy in my study. Fortunately no one can see what I am calling "writing" or they might notice that I eliminate a comma and add a semicolon, add an "ing" to a word and then reconsider it. The process gradually produces a better poem or prose piece, and it surely does keep me happy. Both of us, in our aloneness, are supporting each other's work and choices. Thinking of the way we are together, I begin to remember a quotation I copied in my journal, surely thirty or forty years ago. Instead of searching through hundreds of pages of scribbles, I can now type what little I recall of the wording onto my computer screen and instantly discover that what I imperfectly remember is Rilke's comment that "Love consists in this, that two solitudes protect and touch and greet each other," a line I can't improve.

February 24. I am reading my father's journal this morning; my folks were in Texas and on this day in 1981, George wrote that geese were flying north. As I read, I hear geese honk and glance over my shoulder as a half dozen of the big birds splash and clatter to a landing on Lake Linda.

Time folds; for a dizzy moment I am not sure when I am in time. But when I look down the slope toward the pond frosted with geese, I know *where* I am. Ground truth.

February 25. 30 degrees at four A.M. In my journal for January 1, 1973, I wrote, *"Lots of people want bits and pieces but nobody wants me. I guess I must love myself enough for all."* That was the year I turned forty, finally divorced my first husband, and began to remember who I was, and to discover who I might still be.

In reading these journals, writing about them, am I still trying to understand and evaluate myself? To what end?

February 26. At 5 A.M., it's thirty-three degrees. I've gone back to my father's journals and realize how much of stock-raising information he was giving me. Was he, when his mind was clear, preparing me to run the ranch? I must believe that he was, because he knew I was good at ranching and loved this place. Perhaps he hoped I'd find a husband who could help me. Only his deteriorating mind convinced him to accuse me of "trying to steal my ranch."

He writes that bred heifers shouldn't be wintered with cows; they won't get enough feed because the older cows will shove them aside. Heifers after their first and second calves should get all the feed they want. At one year, he said, calves stop chewing whole oats and just swallow them, thus not getting the food value, so it's a good time to feed rolled oats, which digests more easily. Where did he learn these things?

Driving to town, I always shake my head when I see a rancher who has opened a gate to let his cattle feed themselves among his haystacks, mixing old and young. In such a situation, the weak become weaker because the strong push them away. I can't help thinking like a rancher.

One of my father's strongest rules was never move hay with cheat grass in it anywhere; we fed it where it was cut. Botanists now studying the huge problem of too much cheat grass in the Southwest point out that grazing it heavily actually increases its power. Native grasses take longer to recover from overgrazing, so the invasive newcomers get a head start on recovery and increase their hold on pastureland. So ranchers who overgraze because they can't afford more land or can't afford more taxes will damage their own land, which in turn means their cows get less nourishment which means the rancher makes less money at the sale ring which means he is poorer; it's a vicious cycle. And by overgrazing, they are damaging other resources such as water and wildlife that belong to all of us. Still, most of the public criticizing them don't understand how difficult their choices are.

February 27. Still fascinated by my father's ranching rules. There's much more nourishment in cured grass (both hay and pasture) than in spring grass, he said. So, even when the snow's gone and the grass is greening up, cattle need supplementary feed of cured grass or hay to get full sustenance. Passing the ranchettes all around, I shake my head seeing both cattle and horses turned out to graze pastures all winter and spring, getting thinner and weaker if they don't have enough supplementary feed.

Several neighbors have told me of watching ranchette horses actually starve to death because the residents thought the animals could support themselves on ten acres of grass. The neighbors seemed to think this was funny, since it proves the ignorance of the subdivision folks, but what about the animals' suffering?

February 28. I've tried to forget the book club that doesn't believe *Going Over East* is true, but I wake up at 4 A.M. arguing with them mentally, uselessly. I console myself with the thought that what I have written is as true as I can make it, whether or not everyone believes it. Even if some folks disbelieve me, others would confirm what I've said. The books I've written are part of the history of this ranch and of this part of the grasslands for the past seventy years. The journals of these arid grasslands are written in wind, in blizzards, in the flesh of the cattle they've nourished—and in the truths I've recorded.

Every grass blade, every rock, bristles with white frost this morning. Each thistle hair is thick with white ice; each echinacea seed bristles, so the heads look like powder puffs. Each pine leaf is white.

After their initial leap into the day, the dogs curl deep into the bed and later the couch cushions, noses under their tails. I'm puzzled by thick white loops of frost hanging from the barbed wire fence until I realize they are hairs from cows' tails, invisible until the frost outlines them.

Near sunset, cows bawl for their calves, sounds muffled by fog. A good day; I worked on an essay declaring the healthiest wilderness areas to be ranchers' pastures.

February 29. Every day I record the high and low temperatures here, and compare them to my father's records, creating a snapshot of the real weather in one small part of the grasslands. Ground truth, but unlikely to be entered into a "data stream."

"Charlie" Hasselstrom was born in Sweden on this date; almost a century later, his grandson's son was born. The child often stayed with my parents, and my mother recorded his height every year on the kitchen door molding. Last summer he visited from California and asked to see my mother's kitchen, where he whipped out his smart phone and photographed those forty-year-old pencil marks. Ground truth.

— March —

March 1. The sun comes up below clouds lying in long strips above the horizon. As the glow increases, we see a gaggle of at least forty geese floating on the pond.

Sunrises, sunsets, geese: the little details of our days. Jerry goes to his shop; I go to my computer. We meet at lunch and separate again, happy in our work and lives. "Let there be spaces in your togetherness," advised Kahlil Gibran in *The Prophet,* one of my mother's most treasured books.

Reminded of the advice, I find my mother's worn copy of the book and read a little more of the same: "Love one another but make not a bond of love. Fill each other's cup but drink not from one cup. The oak tree and the cypress grow not in each other's shadow."

I suppose young people think "boring" is *bad*; grownups understand it's paradise. We remember plenty of drama: Jerry was struck by a car while riding his motorcycle; my accidents were mostly with horses. We say our excitement is lunch at the gas station or a Sunday drive.

I think that in these slow, steady, lovely times we are storing up the energy we will likely need to cope with future emergencies. Jane Kenyon's poem "otherwise," posted on my office bulletin board, speaks of days like ours. The final lines are, "But one day, I know, / it will be otherwise."

March 2. "The shavings of your mind are the only rent," hums my skull and can hear Peter, Paul and Mary singing. As a reporter for the college newspaper when the group came to campus, I waited on the grass of the airport runway as their little plane landed. Mary Travers, waterfall of gold hair swinging, leapt out of it and flung her long slender self onto the green grass beside me. I remember nothing of our interview except that she spoke as if I were her friend while I asked questions she'd probably been asked hundreds of times.

Are my journals the shavings of my mind? The rent I've paid for space on the earth?

What are these links we feel with people we've only met? Ever since that day fifty years ago, I've remembered her sprawled on the grass as clearly as some memories of people I truly knew and loved. When Peter, Paul and Mary songs come on the car radio, Jerry and I both sing enthusiastically, the music obliterating the ten years in age between us.

March 3. This morning, raising my bedroom window shades at 6:45, I see movement at the west end of the dam. One, two, three, five coyotes amble along, sniffing the ground, looking over their shoulders, looping and weaving among hummocks of grass. Perhaps they are traveling, searching for mates; it's the season. Fifteen antelope watch from the hillside before skittering out of sight.

Government coyote hunters are flying their deadly planes a dozen miles to the east, killing packs of coyotes; I didn't believe the shooting justified, because I've never seen more than one coyote at a time here. Is this a significant change or an anomaly?

A headline says a recent Chilean earthquake was so powerful it may have shifted the earth's axis. How much is climate shifting the earth every day without our notice? Whatever is happening, it's happening whether we know, or believe in it, or not.

March 4. Forty antelope appeared north of the ranch yard today, grazing and lounging in the grass. One always has its head up, looking around. If one of us coughs as we stand on the deck, heads snap up, the big eyes looking for danger, noses and ears twitching, ready to leap away. Their alarm call is a high-pitched *caw*, higher than that of a crow.

I've been exchanging emails with an older ranch woman from the community who says her husband kept diaries of everything they did from 1952 until the mid-80s when he tired of the job. She kept the diaries for only a few months, but she has thirty-two albums of photographs she took of every phase of their operation, labeled with the dates and events. When someone needs to know how deep the plumbing was placed in 1968, she can find the information in the correct album and diary.

This precise journal system makes the other journals I'm reading look skimpy. What a prize the husband's diaries would be for an historian!

March 5. Dawn. The daytime world comes back as more headlights flash past on the highway. The half-eaten moon hangs sullenly overhead. A cow bawls. Sunlight glints off highway signs. The highest peaks of the hills, still snow-covered, glow orange.

I was happily typing away in my study when I realized that the dogs were barking in a way that was unmistakable: DANGER. Toby, as the alpha male, does all the barking, announcing the UPS truck and visitors. Both dogs were barking as I shot outside the door.

A bull snake was coiled and hissing by a raised bed. The danger to a bull snake's career is in sounding like a rattlesnake, probably to frighten predators. Unfortunately, this evolutionary twist didn't consider humans, who often kill the nonvenomous snake.

I dragged the dogs inside and then, leaf rake in hand, looked closely at the coils. Yes, no buttons on the tail. So I slid the rake under the snake and headed for the pasture. The snake kept coiling and sliding until it was firmly entwined in the leaf rake's tines, with its head raised, cocked and loaded, still hissing. And then it struck.

So much of the reptile was entwined in the rake tines that its fangs didn't even reach the rake handle, let alone my hand.

When I put the snake down, it continued to coil and recoil, getting a 360-degree view as its head made a complete circle. While its body turned, it remained coiled at all times, poised to strike. By making the complete circle, it learned that the human was on one side, and no other threats were visible anywhere else. The hissing slowed as I walked away, but the snake remained coiled, providing a visual lesson: snakes survive by being constantly vigilant, looking in all directions for a threat that may never come. This is too early in the season to see snakes; I wonder if he wintered in some warm spot close to the house.

March 6. The wind is blowing ferociously around the house, rattling the siding, making the windows moan, so at first I think the sound I hear is just another wind sound.

Suddenly the howl rises and breaks off into a keening note. Not the wind. I step cautiously onto the deck and realize the bulls in the pasture below the house are bellowing and moaning. They've discovered the blood pool where, earlier today, we had to shoot and butcher a bull that had broken his leg in one of the traditional battles for dominance among herd bulls. Half the night they bellow and groan. In the morning they gather again where the bull died, pawing, bumping heads, bawling. What does the sound mean? Fear? Sorrow? Folks who think of cows as placid, stupid creatures ought to hear this.

March 7. "You're gonna carry that weight, carry that weight a loooooong time." This morning it's the Beatles singing in my head, one line, over and over. No question about this reference: the weight of my mother and father's journals bows my neck and makes my shoulders ache. What will I do with them when I finish reading them?

March 8. Trying to avoid thoughts about my father today, I turned to a tiny booklet, two inches by four inches, my uncle Harold kept briefly in 1979. Perhaps he'd been listening to my father talk about his daily journals and decided to try it himself. Whenever Harold made a statement like, "This is the coldest January we've ever had," or "the wettest June," my father would consult his weather notes and say, "Well, no, actually according to my records, the coldest year was. . . ." continuing the kinds of disagreements they must have had as competing brothers.

In this brief record, Harold didn't have much space; he noted rain or snowfall and who he saw that day, the work his hired man was doing. He went to a fair meeting; my aunt Josephine had her hair fixed.

On March 6, he wrote, *"Had operation today, real good job,"* an economical statement for what really happened. He'd been walking across the corral when he fell and broke his leg. Doctors suspected a serious problem, but lost the first set of test results. So they pounded a pin into the bone to help set the leg. I remember Jo's shuddering description of the sound that pounding made. Meanwhile, the lab tests were redone and proved the leg was cancerous, so doctors amputated it.

On March 7 Harold wrote he was *"sour as hell,"* probably meaning "sore," but on March 8, *"Not so sour today, will get better fast now."*

On March 10 he wrote, *"Cold back here. Linda George wedding."* I'd forgotten that he couldn't come, but we visited him in the hospital. On March 15, *"Just another day in hospital."* Later, he mentions when they took the stitches out of *"my stump."*

Flipping through the book, I'm surprised and delighted to see how often he said *"a good day,"* or *"nice day."* What a good attitude for a hard-riding man who had just become one-legged and was struggling to learn how to use crutches.

On January 1, 1980, Harold wrote, *"Sun is out today. 30 before it came up,"* and that's the last written word of his in my possession. He lived sixteen years without his leg, learning to swing deftly in and out of his trucks and tractors, always wearing the grin I remember.

What do these few scribbles in this little book mean to me, or to the world?

I'm reminded that my father raged for days when he found out that Harold didn't ask the hospital to give him his diseased leg to be buried in his grave to wait for the rest of him. "Why would you let them just—just throw it away?" he shouted.

Harold wrote every day of 1979, carrying the little book in his shirt pocket; yes, it carries a faint odor of sweat, like dry hay.

Moreover, the book has brought back to me my uncle's face and his laugh, how he talked with a toothpick in the corner of his mouth, looked sideways and grinned when he made a joke. I look up now at the photograph of him on my study wall. I've paid little attention to it since I hung it there beside one of my father, but now I look closely and see him giving me that look from under his bushy eyebrows, his lips just about to smile as he asked me some question about pasture or cattle that would be tough to answer.

March 9. 18 degrees just before sunrise. I woke up remembering my dad said once that both Harold and another neighbor, Paul, had a hard time keeping hired men because they were such perfectionists. My dad said, "I'd sooner work for Harold than Paul, and I'd sooner be in Siberia than work for either of 'em."

March 10. On this day in 1979, I married George. We laughed with our friends, had our picture taken with his son, and invited everyone back to the house afterward for a potluck dinner. The perfect wedding.

Looking at the journals of others has made me think about my own life. I recall little before I was five years old, when my mother left my father and brought me to South Dakota. She persuaded her mother to take care of me in a little house in Rapid City while she worked as a secretary a few blocks away. I learned to roller-skate on a sidewalk beside a busy street. My knees were scarred for years from the times I hit the concrete. We owned nothing but our clothes and a little furniture. My friends and I played Cowboys and Indians around one lilac bush behind our diminutive house. I remember firing my pearl-handled cap pistols from behind the trash barrel in the alley. If I had grown up in that city, I would be an entirely different person even though it is only about twenty-five miles away.

When we moved to the country, my place changed forever. Nervously, I explored creaky old buildings dusty with pigeons and mice and straw. I could reach under the hen with the mean red eyes and take an egg. She might peck my hand but I would survive, and I would not hurt her because tomorrow I'd have to get her egg again.

I could go into the barn, take my saddle down from the rope where it hung safe from mice and struggle to shove it onto the horse—or let my new father help me—and ride out into the distance.

The endless grass taught me that my place was sunlight, coyotes howling at night, antelope slipping down the draw, and cows standing broadside to the sun. The place could kill me; I learned that early. I was not its master, but I could live comfortably and happily in it.

From the beginning, my sense was that I became part of the land, rather than that we owned it. My parents taught me limits that had to do with how I treated them and other people and animals, based on a particular set of ethics and morality that could be summed up as doing right by everyone else. All this learning took place in the rolling grassland, in its weather and sky and among its animals. Prairie was the language that defined me and my being here amazes me still. I might so easily have been placed somewhere else.

March 11. The description of rain I read in my current mystery—how it puddles in city streets, gleams in the darkness, sings in the gutters—makes me recall the sounds it makes here in the country: whispering through the grass, gurgling down the gutters into the rain tank. I miss rain; we are so parched.

Weeks have passed since we've had snow or rain or any moisture and then it was only a few drops. The grass is getting green but the earth between clumps is dusty. Authorities warn that the fire danger is extreme for this time of year, and sparks from machinery have started a number of fires.

Late in the afternoon I visited a ranch woman from this community who comes back to her ranch fairly often, but lives with her daughter in another state. I was interviewing her for a local history, collecting her memories of the county inhabitants when I thought to ask another question.

When she was on the ranch, I asked, did she differentiate between "town clothes" and work clothes?

Oh my! Not only did she agree that there are clothes for town and clothes for ranch work, but she mentioned "church clothes," and clothing for messy ranch jobs. "I still won't wear jeans to shop," she said. "Or shorts."

She's in her nineties and unlikely to change. Twenty years younger, I was raised by a woman of her generation, but I've made compromises. I wear jeans to town, but I'd never wear shorts in public unless on vacation a long way away from here.

She and I recognized one universal rule: if you are fixing a carburetor or mowing hay in greasy work clothes and make a sudden trip to town for parts, it is absolutely inevitable that you will meet the pastor of the local church or a relative who is dressed in her finest.

My mother introduced me to the concept of different clothing for different activities when I lived in town at five years old. I had to get into my "after-school clothes" before I could play Cowboys and Indians in the alley with the neighbor boy. I had dress shoes and lacy dresses that I wore exclusively for church.

Woe would be me if I was told to play while my mother dressed, and I ventured outside! I was, and remain, unable to go outside without pulling a weed, picking up a rock, kneeling to look at a bug or a plant, and when I did so in my dress clothes, my mother's fury was loud, colorful, and usually painful.

As soon as we moved to the country, Mother discovered more nuances in the clothing categories. When I rode horseback, she insisted I wear a broad-brimmed hat to protect the complexion she was sure would help me attract boys, since, she said, I wasn't beautiful. I wore riding boots because ordinary shoes might get caught in a stirrup and I might be injured if the horse bolted. Overshoes covered either work or school shoes when I went outside in inclement weather. I never wore sandals; rattlesnakes could be anywhere.

Now that I'm back on the ranch with Jerry, we have simplified our clothes stratification. Jerry is a retired engineer. In his life as an employee of a state organization, he wore a jacket, dress pants, and a tie to work every day for thirty-five years. On "casual Fridays," he could skip the tie.

So when he retired, Jerry hung two ties with his dress jacket in a bag in the basement, to wear only for funerals. When he's in his wood or blacksmith shop, his work clothes are clearly identifiable by sawdust, grease stains, threadbare spots, and sometimes patches or rips. When he heads for town, he usually puts on a clean tee-shirt and jeans unless we are hauling the garbage in the pickup.

My first requirement in work clothes is comfort, because I sit at a computer in the basement between racing upstairs to do household chores and cook meals. For ordinary trips to town, I usually wear pants or an ankle-length denim skirt, nothing shorter since I'm at that stage of life where I prefer to conceal both my legs and upper arms. For an evening out or a speech, I wear long skirts.

Time and circumstance dictate my gardening wardrobe. I wear denim coveralls with long-sleeved shirts (against thorns, mosquitoes, and flies), tall boots (against rattlesnakes) and a broad-brimmed hat (against sun.) Visitors who arrive in sandals or flip-flops give me nightmares because they have failed to consider rattlesnakes and stickers or the unpleasant side effects of strolling in pastures frequented by cows.

I am not much of a churchgoer, but when attending funerals, I wear a dress, and have been amazed to see women wearing pants, and even jeans or shorts, while some appeared in everything from shorts to coveralls.

What about church, I asked my rancher friend; what does she wear to church?

"It's a matter of respect," she retorted. "I dress up when we go to church. That means I wear a dress. My son-in-law, on the other hand. . . ."

Apparently younger generations view these matters differently.

March 12. At quarter to seven this morning, the sky is black with heavy clouds, with one thin slice of yellow light at the horizon. Fifteen minutes later, the cloud cover has turned deep blue, above a border of lighter blue sky, and the horizon is outlined in gold.

Flipping through an old journal, I caught a paragraph about my fears of going crazy. *"I might reach a height, but I'm always weaving a net of words to catch myself."* A net of words might also serve as a ladder, a way to climb back to sanity.

Here's another reason for keeping journals: the opportunity to look back and see how far one has come from some hell you once thought normal. Reading my journals of bad times in my life reminds me how fortunate I am now; recording those times hasn't allowed them to dominate my life. I look at objects—that vase, this necklace—and can no longer remember who gave them to me. The past, with its happy or awful memories, still seems to shift, no matter how carefully we record its details.

March 13. This morning the shortgrass prairie is shimmering pale green under a lavender sky. I sat in a chair on the south side of the house and watched clouds pass, changing the look of every wrinkle in the hillside every moment, just as a human face changes expressions.

As a rancher, I have become an expert on manure, the substance most folks avoid in reality even if they employ it in slang speech. I've seen manure in every conceivable color and consistency, from the mustard-yellow of sick calves to the thick black of coyote scat. My specialty is cow manure, which I've splashed through in rubber roots and, when I lacked foresight, in every other footwear from sandals to my best riding boots. In the corrals, I've waded in a thick mixture of mud and manure nearly to my knees in spring, trying to get a cow through a gate without falling face down. I've been slapped in the face by a cow's tail slithery with it as I tried to get a calf to suck. I've shoveled it dry into buckets and wheelbarrows and broadcast it over my garden, working it into the soil as I walk and plant and cultivate.

Mostly I am a student and admirer of cow manure. As I walk through the pastures, I automatically kick the big flat disks of dried cow pies, noting the tendrils of grass that have grown under them. I flip each one over if I can and break it into pieces so it will break down faster to nourish the grass instead of smothering it. I always marvel at how cows turn grass into manure and then into grass again. Analyzing cow pies would tell us where those cows were eating; is it too much of a stretch to say manure is a journal?

Is writing the same? We turn ideas into rough drafts and then perhaps into writing that inspires other ideas.

March 14. The first red-winged blackbird sings in the early morning; we've heard no meadowlarks yet. George always contended that the meadowlarks were the real harbingers of spring, so each time I hear the red-wings first, I think of him. A light fog hangs in the air, as if his spirit is hovering nearby to continue the argument.

As the sun comes up, we can see faint trails through the grass: walking or driving even once through the grass leaves enough of an impression to show in this early, intense light. Yet some folks pave over this prairie without ever knowing how delicately it records passages.

March 15. My journal tells me that in 1977, when I was 37 years old and still had choices about having children, I was reading *Passages* by Gail Sheehy, who notes that about four hundred times in her life a woman must decide whether to leave herself open to pregnancy or "deny her uterus its animating powers." Being "free," Sheehy says, "requires an act of negation every month." A woman can't ignore the options.

No wonder that period of my life was difficult. Statistics assured me divorced women were more likely to be hit by a meteor than marry again. My mother talked and behaved wildly, blamed me for destroying my first marriage, and said that by doing so I had denied her right to have grandchildren. Yet she knew my husband had been seeing other women.

I probably never told my mother about meeting the doctor who had performed an abortion for that husband and one of his girlfriends. I was seeing a counselor who told me that my biological father's abandonment made me believe I'll always lose what I love. My history at that time was of driving away men who cared for me, ensuring the prophecy came true. I could blame the cheating alcoholic who sired me, or my first husband, for my insecurity, but it was more important to learn how to stop it.

Perhaps here's a good reason not to keep journals. Still, even if we don't write these devastating incidents down, they remain in our buried memory and still might influence how we behave. Writing painful memories puts them in a form we can analyze and study and perhaps learn to change.

March 16. Today, as I was writing the date and weather in my journal and making notes on what to have for lunch, I noticed what I was doing. Even now that I am an experienced writer and know my journals to be the foundation of much of my writing, the rich soil from which so much of my work has come, I often write daily details rather than the important things that are on my mind.

My first husband and I had been separated for several months when he found my cache of journals and read the most recent ones, trying to discover if I had been seeing another man. I burned them. Over and over, I've told audiences that your journals should be private. If a person will read your journals without your permission, I've repeated often, that person needs to be removed from your life. My journals were a record of my life on the ranch from the time I was nine years old until I was in my mid-twenties. By burning them, I lost all the history I'd written there. When I look back at this act, I can hardly believe I made the choice to take him back after his betrayal.

But have I been mourning a loss that was not serious? Perhaps what I burned were details. Perhaps those journals did not contain any important information about my life.

Still, no matter what was in those pages, I foolishly destroyed the evidence of my life. Today, I understand that the destruction was an overreaction to his infidelity. I can blame no one but myself.

March 17. My memory is never linear, never announces, "On March 17, 1966, you did this. . . ." Instead, it throws up vignettes, like the shifting scenes in a kaleidoscope. Often I write these down at the time because I am afraid that, like the colorful visions, I can never recreate them, never find them again.

Searching through my father's papers has given me a new perspective on the problems he faced then. His tax records for 1988–89 show Custer County had been designated as eligible for federal drought assistance. The ranch usually produced 150 tons of hay and sufficient grass for 180 to 250 cows. The 1988 hay crop was twenty percent of normal. During 1988, we sold our calf crop, but we also sold heifers that we had saved for breeding, and some cows that were the foundation of our herd, a total of 160 head. My father noted that he sold ninety head of cattle he'd have preferred to keep.

The 1989 records show an even bleaker picture. Again the county was designated for assistance, and the 1989 hay crop was approximately ten percent of normal; we sold one hundred head of cattle, including thirty we would normally have kept.

Because I stayed on the ranch after George died in 1988, I knew how dire the drought was, but I didn't understand the finances, because my father refused to discuss them.

My father's journal for August 5, 1990 says he and Mother discussed the changes in the community, how this ranch might look "a few years after we are gone," but I was never part of that conversation.

In 1990, two years after George died, I still believed my father would make me his partner in the ranch, but when I examined his will after his death in 1992, I realized he had left the ranch to my mother, disinherited me even before George died. He had the right idea; after George died, his ex-wife sued me, trying to get half of the ranch.

After my parents left for Texas in December 1990, I had trouble with their kitchen range; a repairman found five gas leaks and other problems with their furnace. Were they inhaling enough propane vapors to harm them mentally and physically?

March 18. Scanning my own journal for 1983, I'm nervous, trying to glance ahead enough to miss the depressing parts, hoping to find something good. The turmoil in our lives was awful: George had a fractured vertebra and a lump in his back he was afraid was a return of Hodgkin's disease. My mother came home from Texas and berated me because some of her plants had died. On a teaching job in a school, I left my journal in my briefcase in the teachers' lounge and came back after class to find a teacher reading it.

Walking to my car that night, I passed an elderly couple on the sidewalk and smiled at the loving picture they made. She spoke softly and clung to his arm; his eyes were blue. As I passed them, he said distinctly, "Up yours." Now I know that he was probably suffering from dementia and she was having to deal with it.

March 19. After a day of snow with sixty mile an hour winds, killdeer chatter along the edges of Lake Linda; where three seagulls float on its surface.

I proved today that a true ranch woman can put on lipstick while driving across the rippled pipes of an auto-gate at thirty miles an hour.

March 20. Being back on the ranch has made me think nostalgically about my horses and recall some of my horse wrecks.

My first horse was Blaze, bought from a shady neighbor without my father's advice; she educated me early. Coming home from checking the pasture, we stopped to water the horses at a corral tank which was leaking; on one side was a deep slew of mud and manure. The horses drank on the dry uphill side, but when Blaze finished, she took two steps forward and dropped to her knees. I yelped in surprise and jumped onto dry ground. Then she rolled, turning completely over on her back and wiggling her legs as she ground my saddle into the mud.

Shaking his head, my father said, "She got too hot for an old fat horse," he said. I spent hours scraping and brushing mud off both Blaze and the saddle.

My dad bought a showy palomino we discovered was a little snortier than his reputation. One day as I was riding him, following my dad into the corral, the horse reared and went over backwards, *whump*!

My dad spun around, looking scared, but I'd blown out of the saddle backwards. I was lying flat on my back gasping for breath while the horse floundered, trying to stand up. If I'd been under the horse and saddle, I'd probably have broken my back or hips but the corral dirt cushioned me. We took the horse to the next sale.

Finally Harold gave us a thoroughbred mare we crossed with an Arab stallion to raise good colts. The first foal from that breeding was my beloved Rebel, who was my primary horse for many years. The second colt was Yankee, gorgeous but spooky. He developed a phobia for white quartz, which is scattered all over our pastures. One day he spun around without warning, nearly unseating me. When I realized he was shying away from a white quartz boulder almost buried in grass, I rode him closer; he bowed his neck and leapt sideways. Finally I dismounted and walked up to it. He bolted, nearly jerking my arm out of its socket. Ultimately, I just had to avoid white rocks. No matter how many times I patted him and walked him up to a big white rock, he never got over it, but we kept him for years.

I suppose I rode a horse several times a week from the time I was nine years old until I was eighteen or so. I miss riding, but I could never ride a horse enough to keep it mentally and physically healthy now, with all the other activities I need to do.

Late in the day, flies bumble. Long skeins of geese waver through the sky, heading north. Twenty antelope nibble at faintly green weeds in the dirt where the dam has dried away. Six geese swim in a spot of open water the size of a swimming pool on the pond.

March 21. This is the vernal equinox, called Eostar by the ancients, and Easter in modern Christian tradition. An old story tells of the earth being hatched, hence our tradition of Easter eggs.

As Jerry walked to the highway mailbox for the newspaper, he saw two bluebirds. He mentioned to them that we have installed nest boxes on the old outhouse and several trees in the yard; bluebirds have never nested there. As usual, they flew away toward the hills instead of toward our nesting sites.

When I stepped outside to cheer up the dogs, I saw the three seagulls flying around Lake Linda, looping low over the geese perched on Goose Island.

On this date in 1991 my father wrote in his journal that every time he tried to call my house at the ranch, he got my "answering service." My mother insisted that the woman who answered my phone wouldn't tell her where I was. When I reminded them that I had an answering machine to take messages when I was away from home, and that the voice on the machine was mine, they seemed to understand, only to make the same complaint the next time they called and I wasn't home.

My book *Land Circle* was about to be published, so I was being hired more often to present workshops and talks in the area. I still spent most of my time doing ranch work; I'd been offered several full-time teaching jobs elsewhere but didn't want to leave my parents entirely without my help.

After I moved to Cheyenne with Jerry in 1992, neither of my parents seemed to realize I was no longer in the house on the hill where George and I had lived for more than ten years. Mother insisted I'd never given her the phone number of my house on the hill and sometimes knocked on the door to ask the renters when I'd be home. Finally I stuck a note to Mother's wall telephone with the phone number of our house in Cheyenne. When Mother called, she'd always say, "You'd better come home. You know your father is always right."

Writing all this down reminds me how entirely insane it was, but I wasn't ignoring the situation. I asked my father's brother and sister what to do, but neither had suggestions. I appealed to their doctors and to my attorney and theirs; all said they could do nothing.

Really, what could anyone have done? To acquaintances, they seemed capable of managing their own affairs.

March 22. As the light grows, I see out the bedroom window a dozen antelope moving slowly over the grass, heads down, grazing on green grass so petite it makes only the faintest of shadows on the hillsides. Red-winged blackbirds sing in a bubbling rhythm, entwined with meadowlark song. I stop suddenly and realize I've been so busy I didn't notice when the first meadowlark appeared.

Reading my father and mother's diaries of the year before my father's death has become too painful, so today I turn to my mother's early journals. On August 3, 1923, when she was fourteen, she wrote, *"I just love to dance. I'd miss every show in the world to dance."* Quite a statement considering her diary for the preceding few weeks listed the shows she'd seen: "Secrets of Paris," Jackie Coogan in "Oliver Twist and the Iris," Harold Lloyd in "A Sailor-Made Man," "Fury" with Dorothy Gish, and something called "Bees in His Bonnet." She wrote, *"I was singing 'Yes we have no bananas,' and took a step back and fell in the rose bush."*

At her age, I spent August on my tractor in the hayfield.

March 23. At Easter dinner I learned that my father's oldest sister, Hazel, was his nurse when he had rheumatic fever. He was bedridden for the winter and she stayed with him and nursed him; no doubt their mother was busy with the other children and ranch work. As a result, Hazel got rheumatic fever and had to go live with a half-sister to be nursed back to health. The fever weakened her heart as well but she didn't know it until late in her life.

When Hazel died, in 1986, my father wrote in his journal only, *"Hazel has left us. Know she is no longer suffering."* Her daughter recalls that he refused to leave the hospital room until the undertaker insisted. When he came home, he drove over east and didn't come back until nearly dark, then refused supper and sat on the back porch. When I realized he was crying, the only time I ever saw him cry, I sat down beside him and put my arm around him. He shrugged away from me and went in the house.

He would not, could not, talk or write about his pain. People who regularly "share" their deepest feelings on the Internet for strangers to read would find this silence unbelievable. And I think it was the fatal flaw of the pioneering generations. They had learned so thoroughly to be stoic that they could not show their love. Over and over they said, "What cannot be changed must be endured," and "Talking doesn't get the work done." And working was what counted.

Sitting in my reading chair with my father's journals, I can look down the hill and see the steps where he cried and where he collapsed into death, dying instantly, just as he'd hoped.

What if Mother and I had known how Hazel had nursed him? I have no idea what it means to have siblings, but my mother did and would have understood him better with that knowledge.

When I was studying Faulkner in college classes, my father read all the books and was particularly taken with a passage about Mink Snopes, who refused to sleep close to the earth because, he said, the minute a man is born out of his mother's body, the earth tries to draw him back down into it. If he lay flat on the ground, Mink said, he could feel the earth pulling him down. My father quoted the passage to me again and again; it never occurred to me that he was afraid of death.

I have neglected my own daily writing because of being immersed—overwhelmed, drowned—in my parents' journals. Every day I read a little in their journals and despair at their lack of understanding of each other, and mine of them.

Every day I have gone out and thrown myself into noticing the red-winged blackbirds, the greening grass, and the signs of returning hope.

I have been struggling to understand my father's frailty and his failures because I always saw him as the ideal to which I aspired. Now I have to admit that he was just as flawed as the rest of us.

If only we could have admitted that to one another. Would we have been happier? Or do we have to keep up these impossible facades in order to function?

March 24. While we were drinking coffee in the bedroom, we heard a single *"Honk!"* from Lake Linda. A flock of seagulls was swooping and diving and screeching around a lone goose. For fifteen minutes the gulls harassed the goose, who floated with apparent serenity. When the gulls flew north up the draw, leaving the goose alone on the dam, it began to honk, and a second goose popped over the bank of the dam. Was it hiding?

I feel as if I have spent the last month digging ditches as I wrestled with my father's journals. Will I be able to put him to rest now? Can I accept his great guidance and understand that he loved me even though he couldn't say so?

He would hate that I have written poems about him and recited them all over the country. When I recite them I can hear his voice but no matter how truly I write, they are not the truth. Still, they are all the knowledge of him that is left.

March 25. My journals again. In 1983 I was working on an essay about an epic journey returning from a job in Lincoln, Nebraska, that wasn't published until 2002. In June 1983, I wrote the draft of a poem published in *Dirt Songs* in 2011. This is the life of a writer, I realize, or perhaps of anyone who keeps journals and in them reflects on life. Write and write and think and revise. Perhaps in ten years, or fifteen, writing that will speak to someone else is accepted for publication. No wonder writers and other artists are insane.

March 26. At sunrise, every blade of grass stands outlined in light, a glorious morning.

In this modern day of instant writing and publication of the freshest thoughts, how is a writer ever to have the time to reflect and revise and improve the writing? Very few bits of writing are worth someone else's time to read before they have been polished. Millions of people spew forth their ideas into the Internet ether every day and are praised for this "writing."

March 27. Tired of reading words, I look at objects in my office that form a different kind of journal. What souvenirs of love or escape have I kept that will puzzle those who clean up after me? My spurs, and my father's, hang on the wall above my computer, clear symbols. The medicine pouch I wore at rendezvous contains tiny souvenirs from that life that only I understand.

Some signs are clear: a string of pins from my attendance at the National Cowboy Poetry Gathering in Elko, Nevada. Others may baffle: why is this ugly mirror with the broken frame hanging above my computer? What about the piece of wire that looks like a cowboy wearing a hat? The owl skull? (Someone shot the owls and hung them on the fence. I kept the skulls to honor them). The toy stuffed Westie puppy? (I bought it for George when he was dying in the hospital because I couldn't bring in our real dog.)

March 28. My father wrote in 1981 that he heard a crow on this date and a killdeer yesterday and that the meadowlarks were *"making a lot of noise,"* so he knew spring was coming. I'm trying to imagine how he felt knowing that he faced another year of the same welcome work. Then I remember him looking in the mirror once when he was at least eighty and saying, "You know, I don't feel any different than I did when I was sixteen." At the time I couldn't understand. These days, when I look in the mirror, I do.

March 29. A doe and her yearling fawn graze at the west end of the dam where minuscule flat leaves are growing green.

Jane Siberry sings:

> "I'm bound by the fire
> I'm bound by the beauty,
> I'm bound by desire,
> I'm bound by the duty."

That could be the song of a ranch woman. Then the chorus:

> "I'm coming back in five hundred years
> And the first thing I'm gonna do
> When I get back here is to see
> These things I love
> AND THEY'D BETTER BE HERE!"

I wonder if I could find a recording of that to play at my funeral.

March 30. Snow falls heavily; I-90 is closed and schools are closing. The pond is nearly frozen over again. I hope the red-winged blackbirds and meadowlarks are sheltered in the windbreak trees and haystacks. During one blizzard I pulled down a hay bale and from behind it flew a dozen meadowlarks.

Walking past the barn, we hear a constant cooing from pigeons making a mess of feathers and droppings, and we begin a discussion we've had before: what should we do with the barn? A classic shape with a mansard roof, it has been mostly unused for years; the windows are out, and holes dot the roof. We can't justify the labor and cost of stabilizing or repairing it, or even tearing it down to build a useful shed. Jerry keeps our tractor and its attachments on the lower floor. The shelves may hold a few ancient horse-grooming tools, but otherwise the barn shelters only rodents, pigeons and memories.

Once inside, I remember my father buying a brown-eyed Jersey cow so I would have the experience of learning to milk. I loved dipping an inch of cream off the top of the milk and how that thick cream tasted on cereal or chocolate cake.

When a cow had trouble calving, we'd tie her head to one post, then tie a hind leg back to another stout upright. My father would take the calf puller that's still propped up in a feed bin and move up close behind her. He'd reach inside the cow, laying his cheek against her hip bone, until he could pull one or two feet of the unborn calf out where I could attach the chains to them and to the puller. All this time the cow would bawl and throw herself from side to side, kicking up dust and usually squirting digested grass out the appropriate aperture, just above the birth canal. Then I'd get the crank and we'd begin slowly winching the calf out, trading off as each of us grew tired. When the cow struggled, she'd swing the long arm of the calf puller from side to side, occasionally knocking one of us down, but we'd get up and dive back into the battle.

With a whoosh and a plop, the calf would drop out and hit the ground. My father would put the calf puller back into the feed bin while I grabbed a gunny sack and rubbed the calf. I'd stick my fingers down its throat to make sure the tongue was out, and clear its nostrils with my fingers, sometimes pushing down on its chest until it was breathing. Usually the cow would stand quiet for a few moments, exhausted, while my father untied her hind leg. Once I was out of the way, he'd cautiously begin to loosen the rope and hope that the cow didn't charge one of us. Usually she'd turn and begin to lick her calf, and we could look at each other and laugh and lean on the fence until we saw the calf get up and stagger to the right spot and grab a teat. When the sound of slurping filled the barn, we'd start wiping blood and fluids off ourselves and head for the house.

Some of the square posts holding the second floor up show marks of the ax that shaped them. Others have been gnawed nearly round by the ropes we tied to them to hold the struggling cows. The smell of our sweat seems to hang in the air. If I sat quietly inside, I'd hear cows bawl and smell dust, feel a cow's head slam into my hip as she boosted me over a fence.

We consider having the local fire department burn the barn for practice, but for now we'll let it stand, a safe containing my memories.

March 31. A heron stands on the pond, looking like a Chinese painting through the fog. I love seeing them fly because they carry their necks in an S shape and trail their long legs.

We often see two great blues standing on opposite sides of Lake Linda hour after hour, patiently waiting for the perfect strike. When the great beak descends and then flashes backward, we know the bird is swallowing prey, usually a frog or small turtle, though it may also eat snakes, insects, and smaller birds. Sometimes we sit quietly on the deck watching the darkness flow over the prairie until the herons are completely invisible to us. Yet even in the dark, the birds continue to stand, hunting, because their eyes allow them to see so well. Even in darkness, the small prey is not safe from the mighty yellow bill.

I started reading about Vipssana meditation, about gurus and yoga, but stopped when I realized that the prairie is both my path to enlightenment and enlightenment itself.

The prairie is suitably immense in age and serenity; its silence and depth provide a meditation room for everyone who pays attention. Sit in silence and open yourself, say the gurus, and you will see god or goddess in whatever manifestation most appeals to you.

Around me, I see nothing that is not holy. Goddess in the grass, the birds, the clouds, antelope, muddy pond. Goddess in the sunlight.

— April —

April 1. On April Fools' Day, my father always found a way to fool my mother and me. "Did you see what's in the back yard this morning?" he'd say, and we'd rush to look.

On this day in 1981 he wrote in his journal, *"The railroad started a fire east of tract* [railroad track] *at the crossing at Urban."*

Recently I read bits of a gigantic local history book put together in the 1970s by some local men and women, so I know that when my father was growing up near here, Urban was a community of ranches. Hermosa was a different town, with churches and stores.

When my mother married John Hasselstrom, he was living on what everyone referred to as "the old Sharpe place." Jonathan Sharpe had emigrated from Cumberland in England. Besides homesteading, he worked on the railroad and as a logger in the hills, as did most homesteaders since the land would not support them.

According to the local history, he abandoned his first home site, said to be "on a hillside facing east." I now believe that home site was near where my father found a shallow rock-lined well, east of the ranch house on the edge of an alfalfa field. Because it was a danger for livestock and people, we tossed rocks down the well to fill it. If Sharpe's homestead was in the rich bottomland beside the well, he may have abandoned it after a flood like ones we later saw in the area.

In 1912 Sharpe moved to the site where my retreat house now stands and built a handsome two-story house. Forty years later, father was "batching" there when he met and married my mother in 1952. They hauled Sharpe's house into the hayfield, set it on discarded railroad ties, and we moved into it for the summer, while our new house was being built among the trees

Sharpe had planted, north of the barn and corrals he'd built.

I loved that creaky old house because for the first time, I had my own bedroom upstairs; I slept in the corner of my mother's room in our rented house in town. I even had a can with a toilet seat on it so I didn't have to go to the outhouse in the dark. My father kept the path to the outhouse mowed, but none of us wanted to venture out there among the rattlesnakes at night. I didn't even mind emptying the can into the outhouse every day.

All that summer, I could stare out lofty windows to watch golden boards frame the new house or look over the green, rolling hills while I dreamed of a black stallion.

Until this moment, I had not thought to admire the way my mother adapted to living in that house with no plumbing after years of town life. We hauled water; I took baths in a washtub in the kitchen that she filled with hot water from a tea kettle. She cooked on a wood stove, and another of my chores was bringing in wood. What a change from town! But perhaps Mother recalled growing up on the Hey ranch well enough to adapt, and probably looking forward to having her own house made the experience bearable.

The local history also says Jonathan Sharpe gave land for the first schoolhouse in the Urban neighborhood. When I moved here at age nine, an old schoolhouse had been converted to a granary. My mother's journal for 1957 reminded me that I was fourteen years old when I helped my father tear it down. I remember being perched on a rafter, inexpertly pulling nails while he reminded me of the textbooks my mother had rescued from the building's mice.

Not long ago, walking the dogs, I realized I was walking along a half-buried line of dressed sandstone block west of the foundation of the old school house. Jerry and I poked and prodded, following the line of stones, and decided they marked the edge of a road in front of the schoolhouse. Friends brought a metal detector, and we unearthed two decorative door hinges and an oil can. Every few days since then we find broken window glass, fragments of pottery, shingle nails, or horseshoe nails.

I can picture the classroom where the children in the Urban community learned their lessons, warmed by a pot-bellied wood stove. They probably put their horses in a barn during classes or tied them in the corral we still use. Only the square-cut stones of the foundation remain, part of the unwritten journal of this place, laid down before we came here.

If Sharpe was the first white settler here, this ranch is more than a hundred years old, but on the journal of the land, the English settler's marks are faint as pencil on an ancient page.

Horned larks twitter like voices at a distance, like ancestors trying to tell us something we can't quite hear.

In the dust at my feet I see a coyote track: another journal we can barely read.

April 2. Ten, twenty feet beneath our feet and the hooves of the prong-horns and cows lies an intricate system, the interwoven roots of native prairie grasses, stirring now, sensing the snow melting above.

If a scientist discovered and named this incredible organization, and called a news conference, reporters would arrive from every corner of the world. Citizens would revere prairie as they do the Arctic and the Amazon. Yet many people think they are making a wise investment when they cover this grass with buildings, asphalt, and plastic merchandise.

I have no written records from Jonathan Sharpe, but I can read part of his journal in the trees that protect the house. They stand along a streambed that floods only in the wettest years; perhaps he planted them during one of those periods. Even in the dry streambed's sandy, rocky soil the tree roots plunged deep enough to water themselves. Cottonwoods live from eighty to a hundred years. If he planted these trees as soon as possible after moving to this spot a hundred years ago, they are near the end of their life span; I should be planting more. In their whitened trunks and branches we can read his understanding of the need for shelter from the north wind.

For twenty-five years, Sharpe and his wife Elizabeth "diversified their income" in the lingo of today's financial advisors. They raised cattle, farmed and ran a dairy. Lizzie kept chickens to provide her family with eggs and meat. If she had extra eggs she sold them at the local store along with the milk from the dairy herd. In 1938, Jonathan and Lizzie sold their cattle and machinery to the Hasselstrom Brothers. Mrs. Sharpe died in 1940 and the next year Jonathan Sharpe sold the land to the Hasselstrom Brothers and moved to town. They had no children. I am as close to an heir as Jonathan and Lizzie will ever have.

Abruptly, I recall one possession the couple must have left here: *The Poetical Works of Sir Walter Scott, Bart.* The edition is undated but its pages

are brown and foxed, rimmed with gold. Inside the front cover is a slip of paper inscribed in fading ink: *Miss Lizzie Collings*
Taylorville, Ind.
Dec, 1889

Between the pages are fragments of pressed flowers, a four-leaf clover, and a blue ribbon embroidered in ornate script with the initials "EAC." I've found no markings on any of the poems, but at the very least, I know that Miss Collings was sufficiently well-educated to appreciate this lushly descriptive poetry and to pack this 659-page book when she moved to the wild Dakota plains. My mother and Miss Collings, or Mrs. Sharpe, might have delighted in an afternoon together, reading Scott's poems had they known one another. I feel a little better acquainted with Lizzie myself, because I've read a lot of Scott's work.

Walking around the ranch, I often feel as though my father or mother, or George, has just stepped out of sight. Now that I know more about Jonathan and Lizzie, I feel their presence here as well.

My journals and those of the others who lived here, both written and built on the land, are woven into its history. When a neighbor convinced my father to level an alfalfa field by flattening and destroying the shallow ditches Jonathan Sharpe built with his team of horses to spread heavy rain gently over the field, he was destroying part of the web of husbandry that had been woven here. The neighbor saw the field only as a flat place to be plowed, harrowed (I think it is fitting that "to harrow" also means "to inflict stress or suffering on"), fertilized, and made to produce more.

Under my ownership, the field has been returned to native grasses that don't require any machinery for their sustenance. The cattle graze there in the winter, spreading their manure naturally over the field. Rick disks the field lightly each spring to break up the cow pies and help the ground absorb that natural fertilizer. Each time it rains, I wish the spreader ditches were still there.

April 3. After breakfast, as I was looking out the kitchen window, the first buzzard I've seen this year suddenly shot up from behind the hill, making me gasp with the sudden appearance, the black wings slicing the sky. Coasting low over the windbreak trees, it tilted those great wings in one smooth curve. I wonder if the bird is scenting the dead mice we've tossed out into the grass after the dogs have killed them.

Driving to the post office, I watch Hermosa growing in my windshield. Ranches surround these western Dakota towns like sheltering arms, cradling the little communities of homes, a post office, and churches. On winter days, pickups come from all directions as ranchers get mail, buy milk, exchange greetings, put a little cash into the town—even though, sadly, many of us spend most of our money in the big-box stores in the city.

April 4. In late afternoon, we were strolling with the dogs on the hillside as usual when we glanced south toward the stock dam.

At the edge of the water, a red-tailed hawk was thrashing, great wings spread out on the surface. Every now and then he'd flap his wings and try to rise, but was dragged down and rolled over into the water.

Several ducks were gathered in a half circle out in the water, watching. Had the hawk attacked a duck and was having a hard time subduing it? Several times the hawk was completely underwater, and several times it tried to fly but seemed to be pulled back, ducked under, and rolled, wings thrashing.

Finally the bird broke loose and flew to the top of the dam, alighted for only a moment, then rose again and sailed off east, with nothing in its talons.

We climbed through the fence and walked to the spot; the water was cloudy, but among the large rocks I could see no disturbance. No talon marks in the mud; no blood.

As we started back, walking slowly, many frogs leaped into the water. Suddenly a portion of the bank moved, turned its head, and hissed at me: a bull snake. The snake was at least five feet long, the body as big around as my arm, and moving slowly, without the quick slither I expect. As I followed the snake, snapping pictures, it faced me, inhaling and exhaling, coiling and retreating until it backed into the water.

I believe the bull snake was hunting frogs when the hawk struck, but failed to make a killing stroke. The snake was too heavy for the hawk to lift, so they struggled to a draw. How lucky am I to see such things?

April 5. We heard cranes tonight, saw a huge W of them high against the clouds.

I love this quiet time at the end of the day. My bedroom is warm from the western sun but becomes shadowed as the sun sets. The dogs curl sleeping on the bed, pink ears up in case something interesting happens. I love to look at the familiar furniture. Jerry made a table that looks as if it is dancing, with legs of twigs from a tree someone backed over in a store parking lot in town. I watch the sunset from the rocking chair in which my grandmother may have rocked my mother.

On a rack Jerry made hang family quilts, all created from scraps that some women might throw away. At the top is a star quilt given to me by a widowed friend. Lakota custom dictates that a year after the loved one's death, the bereaved give away his or her most valued possessions. The sacred star quilt that covered the coffin is given to the friend who has been the greatest help during the mourning year. The honor she gave reminded me how blessed it is to give, and not just to receive.

Another quilt recalls a job my uncle Bud had during the 1930s. He was given a book full of wool swatches and suit designs, and tried to convince customers to order tailored suits. After he quit the job, my thrifty grandmother Cora made a quilt from his samples. My afghan is vivid blue and green; Grandmother crocheted one for each of her grandchildren. The bottom quilt is a Grandmother's Flower Garden my mother cut out while she was pregnant with me, using bits of her own dresses and Cora's. Thirty-some years later I found the pieces in her closet and a local Ladies' Aid finished the quilting, so these tiny stitches memorialize some of the most experienced quilters in our community at that time. They're all dead now, tucked into the Highland Park Cemetery near Hermosa. This quilt and the others are part of the journals of those women whose arthritic fingers stitched messages in colored cloth, journals we can read only imperfectly. Every stitch echoes with their laughter and their voices as they worked, discussing and settling community questions while they created this beauty.

Tucked into the spaces between the quilts are other stories. A quilted pillow in a geometric black and red design was made by a Hmong woman displaced by the Vietnam war; I love to recall buying it in Pike Place Market in Seattle. I could not afford the huge quilted pillows that depicted the maker's entire village, but a petite woman with a brown and wrinkled face handed me the tiny one. She has found a way to survive in a new and strange land, and I cherish her face as a reminder of my own ancestors, women who sewed their way to peace of mind on this Western frontier. My pillow is part of the Hmong woman's journal of life in America, sold to a stranger in a state she probably will never visit.

April 6. A mountain bluebird sits on a crossbar under the deck while a robin proclaims the garage roof as his territory.

My father wrote in his 1981 journal that George and I planted potatoes on this date. Last year I didn't plant them until April 21 because we were getting so much rain the garden was sodden.

I've been zipping through some depressing parts of my own journals. Today I woke remembering the strong voice of Meridel LeSeuer saying to me, "I covered some terrible wounds with lyricism." So have I, and most women. Some covered the wounds with smiles painted in makeup; some hid the wounds with threads of embroidery, and some with poetic language.

The Hmong pillow reminds me that when The Moving Wall, a half-size replica of the Vietnam Veterans Memorial, came to South Dakota a few years ago, I was amazed to find myself standing before it with tears streaming down my face, tears for all of them, for all we lost during that time. Everyone else was wandering along the wall, finding names, speaking quietly.

Then I stood back and looked at the soldiers guarding the wall. They were all Lakota Indian men wearing the uniforms in which they had fought in Vietnam. The Lakota have had a difficult time with alcoholism, and these men showed the story: hung-over bellies, beat-up faces, scars. But every damn one of them was standing as straight as he could manage to be, at parade rest. They stayed there for hours, in the sun, watching as mostly white civilians ambled past that wall.

They *knew* what that wall stood for. They built that wall with the men and women who didn't come back. The rest of us were just tourists.

I did the only thing I could think of to do to show my appreciation: started at one end of the wall and walked along facing the men. I looked every one of them in the eye and nodded. I couldn't salute; I have never been in the military. But they understood what it meant that a white woman most likely from South Dakota, the most anti-Indian state in the union by many standards, meant by the action. By the time I got halfway down the line, I was crying so hard I almost bumped into the oblivious folks walking straight to the wall. Some of the Lakota nodded back at me. Others made no acknowledgment. The ones who didn't come back; that's one tragedy. The ones who came back are a whole different set of tragedies. Those who came back to a poverty-plagued reservation in South Dakota surrounded by racism is a disaster we must eventually face.

April 7. We got up at four to see the eclipse, and were able to watch it until the moon was down to a sliver, then the clouds moved in and we didn't see the end.

I shoveled rich garden dirt into my largest pots and saturated them with water for tomato plants. I'll put other plants in the garden below the hill and in the raised beds, but these will be my insurance tomatoes, sheltered from hail by the deck on the south side of the house.

Meadowlarks, killdeer, red-winged blackbirds sing. My father wrote in his journal on this date in 1979, *"The air is full of bird music."*

April 8. After a high of sixty degrees yesterday, sunrise brought thirty degrees and six inches of snow; glad I didn't plant any tomatoes. The cows draw a curling line down the hill toward feed.

On the hilltop, we saw two shapes we didn't recognize as the usual cows or antelope so we each grabbed binoculars and saw two big red-tailed hawks sitting side by side. Every now and then one would fly, circle, and then return, to sit by the other. After a while they both flew low over us.

At sunset, three robins were lined up on the railing of the deck, twittering furiously. Arguing? Exchanging news about the location of the fattest worms? Reciting poems? Declaiming oral journals?

April 9. At twenty minutes after eight in the evening, when I can barely make out fence posts close to the house through the darkness, a snipe makes the call that makes my spine tingle. For years, standing on the deck at dusk, I'd hear the snipe's breathy *whooooo* and asserted that it must fly in circles around the house crying just to drive me wild because I didn't know what it was and never caught a glimpse of it. Once someone identified it for me, I realized I see this common bird often, but have never heard its peculiar sound except at night.

Before Sharpe settled here, assuming he was the first white settler on this land, others left their mark. For years we have marveled over the stone circles in the pasture to the northwest of our house. We thought they were tipi rings, circles of stone used to hold down prehistoric shelters for natives who camped here. So we invited several archaeologists here, and they supervised informal digs in the circles. We dug dirt, sifted it, and probed gently in and around the circles. Sweating, we joked about finding the campsite of the lost tribe of Tidy Indians because we found not one flake of flint, not one charcoal layer indicating fire pits, nothing. Still, the circles are there.

Recently, a historian visited to look at this and other sites and pronounced them prehistoric, representing, perhaps, turtle effigies. On top of the ridge we named for the resident badgers, he examined the line of piles of rock that my father thought was an old wagon road and pronounced it very old, possibly prehistoric as well.

These prehistoric inhabitants left their journals in stone. Perhaps we will someday be able to understand a little of their message.

April 10. The grass around the house undulates and heaves with meadowlarks and blackbirds high-stepping as they catch bugs. Beaks full, they leap into the air and zing away. Sometimes it seems as if the air is a tapestry woven of birds flying fast, busy at their work. The meadowlarks are probably plaiting their grassy nests already, perhaps laying eggs. My usual sources are silent about how early this may happen, though they note that each male feeds two females who build the nests alone.

April 11. At noon, five buzzards hang against the blue, tip and swoop between the winds. They're back! Now they will begin clearing the plains of the animals that have died during the winter, and patrolling along the highways for the fresh corpses drivers will make for them.

My dad noted in his journal for 1989 that I went to Rapid City on this day to see my attorney about George's estate. I thought I was losing my mind because I had not been able to find his original will in the safe deposit box where I clearly remembered putting it; I didn't know then that I should have left a copy with the attorney.

Settling George's meager affairs became a time and money-consuming matter, complicated by the fact that his ex-wife sued me, saying that her son with George was entitled to half the ranch. Over and over during the next few months, I asked my father if he'd taken the will from the safe deposit box we shared, until his denials grew angry. After his death four years later, I found the will in a garbage can in the cellar, with other important papers, like land deeds. He'd been losing his memory and his logical thought at least four years before he died, or before I realized.

The night sky is clear, the stars' shimmer amazing us as it does every night, as long as we look south and east, away from the town and the neighbors with their glaring orange lights announcing their importance and their fear of the dark.

April 12. The wind has blown forty to sixty miles an hour all day, alternating rain and snow, creating wind chills far below zero. Cows huddle in the corral; if they drifted before the wind, they might pile up against a fence to die as they did in 1949, noses frozen to the ground, slowly smothering in their own frozen breath.

When I first step into below-zero wind from a warm house, my mind instantly clears. All thoughts of the future, of philosophical matters, even of what I might make for dinner, slide away, replaced by alertness. I breathe in but try not to breathe too deeply, feeling the air like fire in my lungs as my nose hairs freeze.

A distant sound—a truck on the highway—roars like thunder through the thin air. I open my eyes wide, then squint because the air feels as if it is freezing my eyeballs. I am alone here for a few days. I walk carefully on the ice, knowing if I slip and fall, break my leg or knock myself out, I might freeze to death if I couldn't get back to the house.

My lips feel dry, as if they are cracking; tears blur my sight; I seem to feel ice forming in my nose and throat. My feet are clumsy in heavy boots and within a few minutes I am flexing my toes as they chill, closing and opening my hands to keep the blood moving.

Perhaps I should advertise this experience as mental therapy, healing for the heart, or for any number of physical and mental ailments. Nothing like a real survival test to sharpen one's wits.

My journal reminds me that on this night in 1981 we saw the Northern Lights as red as blood in a vast circle in the sky. Father said his mother always told him red Northern Lights called the "blood aurora" meant war.

April 13. A few days ago the air tasted green, and we saw geese headed north. Today geese hurtle south in freezing winds and sleet.

I'm reading an article in *Smithsonian* magazine about prairie voles. Scientists have just begun to study two types, prairie and meadow voles. The former mate for life; the latter are promiscuous. According to my resource books, we probably have both types here, but it's the prairie voles we notice most often. They clip grass to create a network of trails two inches wide and as much as eighty feet long, winding among rocks and hummocks of brush. The dogs love to race along, noses to the ground, until they find a tunnel that leads to the underground chambers.

The burrows aren't what interest the neuroscientists, however. Prairie voles, unlike 97 percent of mammals, form bonds that last long after mating. Scientists are studying them as a part of learning more about human problems such as infidelity; one notes that the animal's name contains the word "love."

I have no way to judge the validity of the research. What frightens me is that it is being done on a creature that lives comfortably on this place with me and that not one in ten thousand people would recognize if it climbed into their laps. Yet many of those people feel comfortable pronouncing this prairie as worth less than, say, a cornfield. As we subdivide and plow and plant and whizz down the highways on this prairie land, we need to ask what secrets it may hold.

Have we paved over a plant that might have been an antidote to alcoholism or a cure for cancer? What knowledge might we lose when we make uninformed choices about the use of land? "Here we'll put our Big Box Store," we say. "We need one on this side of town so people don't have to drive more than a mile to find cheap plastic objects made in another country."

April 14. So many blackbirds are singing around the house it's like being inside a choir. We saw a gold-headed blackbird today, adding the bass notes: *scrawk!* The enthusiastic member of the chorus who cannot carry a tune.

Before I started reading my father's and mother's journals, and re-reading mine, I could look back at my childhood and see it as all one lovely day glowing in the sunlight of nostalgia. I worked cattle with my father while my mother fixed meals and cleaned house, trying to interest me in those activities. They went to dances on Saturday nights, took me to 4-H meetings, and visited friends in the community. We were a happy ranching family. As a child, I saw my world as stable, supported by a web of friends, relief from the confusion of my early life. It would be easy to be poetically positive about those days, but I know some of my nostalgia is false. The world was fluid and shifting even then. History is always moving over landscape, though its traces may be slight.

Do the uncertainties matter if I didn't know about them? I grew up with parents who cared about what kind of person I became and who set high standards for me. Yes, that means I still expect to hear them telling me what I'm doing that's not quite right. Perhaps that's my problem, not theirs.

The writers who come to the retreat house now are staying where we lived during those years I thought were golden. They all remark on the peacefulness of the house and the beauty of its setting. One or two women have said they felt my mother's benign presence in the bedroom where she slept for fifty years. I'm delighted; I know her nights weren't always happy, but apparently if her shade lingers, she is happy for the company I've placed in her rooms. Perhaps she, too, has returned in spirit to the happy early days. Maybe she's having her own retreat, reading as much as she wants.

No one has mentioned encountering my father's furious shade. Perhaps in death he, too, is calm and relaxed.

April 15. At dawn, the hillside and sky are gray. A dozen cows lie together in one black mass. One stands up and faces east. The sun bulges up from the horizon, molten red-gold. The standing cow stretches and turns broadside to its heat.

Reading my mother's journals has been harder in several ways than reading my father's. I always felt guilty when she was alive, as if I didn't do enough for her even though I knew from an early age that one of her best weapons against the world was guilt. I didn't see the world as she did, but I should not feel guilty, even though it's a nearly-universal feeling between parents and children. My journals during the eighties, when I was back on the ranch after my first marriage, show that I took her to art shows, concerts, poetry readings, lunch, and dozens of other events she enjoyed and at which she was happy to introduce "my daughter."

Once, laughing, she told me a friend had asked her, "Who is that woman with the pointy-toed shoes?"

Mother said to me, "I had to tell her 'God help me, that's my daughter.'"

I've always been boringly conservative in my clothing and hairstyles but she was sincerely embarrassed by me much of the time. When I appeared at her door to drive her somewhere, she'd say, "You're not going out wearing *that,* are you?" Or "When are you going to do something with that *hair?*"

She's been dead more than a decade and every time I get dressed to go out, I hear her voice and look nervously in the mirror.

Do our parents ever lose their power to make us feel worthless? And parents, why, when you have such power, wouldn't you use it to make your children feel strong? A moot question, perhaps, since my children weren't born from my body. I don't believe, though, I've ever said such a thing to my stepchildren.

How did these differences between my mother and me arise? Surely before she married John Hasselstrom, she was my entire world. In her photos when I was four years old, I look like a porcelain doll dressed in ruffles and hats, carrying my little purse with care.

The minute I moved to the ranch, all that changed; I idolized my father and lived in blue jeans and my cousin's castoff Western shirts. Was my mother's anger pure jealousy? I hadn't subtracted her from my affection, just opened my heart to a father I'd never known. But I changed: I wasn't content with ruffles when I could have blue jeans, or patent leather when I loved riding boots. And a horse.

April 16. Three blizzards in ten days have dumped snow in winds twenty to sixty miles an hour. Between four-foot drifts, the ground is almost dry in some spots, and covered with a moving carpet of birds: thrashers, flickers, starlings, blackbirds, meadowlarks, gobbling seeds and insects. Tall grass lies in brown streaks pointing north and south, blown flat, and some snowdrifts show long lines: the tracks of windblown tumbleweeds. Every fence is stacked five feet high with snow-packed tumbleweeds pulling down the wires. I'm so glad to be snug in the house. Thirty years ago today, when the thermometer reached 74 degrees, George and I chose this spot for our house; we walked over here with my folks in the evening to imagine what it would be like to live here. We broke ground on May 11 that year.

Perhaps there was more to my mother's anger with me than jealousy. When I moved to the ranch, I was immediately told that I was part of the work force and given chores: feeding the chickens, gathering the eggs even if the hens pecked me. I didn't feel abused because I had to work; I understood that I was working because I was part of the family and therefore responsible within it. I felt needed.

By contrast, when she was a teenager, Mother often wrote in her journal that she "*did nothing*" or went to a movie. "*July 2, 1923: I didn't do much today, just grumbled because there was nothing to do. I dressed up and played some and had some fun.*"

In 1924, at age fifteen, she wrote that her mother, Cora, gave her fifty cents, her week's allowance, and "*I got my new Jacquette this morning and went down town this afternoon and had a coco cola in Bacon's all by myself.*" She usually adds that her mother and her friend Mrs. Cooper were doing laundry or cleaning house. At seventeen, Mother was on the go: for a ride in someone's car, to the drugstore, to a neighbor's or to a movie because the first talking pictures had just appeared.

I find myself shaking my head in disgust at such frivolity; my teenage life certainly wasn't filled with movies and dances. I can't imagine my sensible grandmother not requiring her daughter to work, but perhaps Cora wasn't as practical when she was younger. All three of her other children were boys; maybe they worked and her daughter didn't. How completely we change and misunderstand each other. From the way she behaved when I was a teenager, I never would have known my mother was so carefree at the same age. Even when I was sixteen I would have thought her romanticism silly, but she lived in a different age.

I am stunned to realize I have learned in these journals that when Mother was writing, they lived in town! Because my grandmother ended her days on the Hey ranch where my mother's last surviving half-brother still lives, I'd pictured my mother growing up there.

When my grandmother Cora was widowed by the train that beheaded my grandfather, she moved to Edgemont and opened a small café with her sister Pearl. There she met her second husband, Walt Hey, who was apparently a jack-of-all-trades, an auto mechanic who also had the skill to rewire the high school.

Now I recall a story about Cora walking from Edgemont to the ranch, at least ten miles. Perhaps she stayed in town with the children of her first marriage, my mother and her brother Bud, while Walt began building a better house in the country. Later, Walt fathered two sons, George and Harry, who did grow up on the ranch.

Oddly, I never thought to ask my mother if she grew up on the ranch; I thought I knew. She never thought to mention it; she assumed I knew. No wonder she could go to so many movies.

Growing up in town would partly explain why she was so adamantly against my love of the outdoors and my father's insistence that I become a working member of our ranch family. I loved to work outside and loathed housework. Mother probably never worked outside or with her stepfather; she wanted me to be a lady and learn the skills of a housewife so that I'd be ready when I married that doctor or lawyer. She gave me a nail care kit once for my birthday; perhaps that was when I resolved to be different from her.

April 17. People say they feel the weight of centuries of human endeavor when they visit ancient castles or crumbled walls where oracles once held forth. But when those standing stones were erected, when the Romans were building roads and the Egyptians were piling pyramids, these grass-covered hills probably looked as they do now. These grasses were surviving, nurturing each other. The continued survival of this grass is a testament to the care of all who have lived by its bounty, in spite of harsh weather.

Reading my journal, I remember how dry it was in 1981: dust hung over the hills like smoke every day; we'd had no rain all spring.

April 18. Low of thirty degrees, high forty degrees; an inch of wet snow melted by noon.

In the tree row north of the retreat house we saw a great horned owl watching us until we got too close. Below the tree were two feet and two wings from a robin, the remains of the owl's breakfast.

Today I found a ledger Mother had reused as a journal while she was in high school. In it she copied poems in green ink and ornate, looping handwriting. In her nineties, she could still recite Edgar Allan Poe's gloomy "Annabel Lee" and "The Children's Hour" by Longfellow. Peeking between the poems are photos of heartthrobs like Rudolph Valentino, but also pictures of Einstein, Abraham Lincoln, Robert Louis Stevenson. Only one unidentified woman is pictured, a dark-haired ingénue with a lacy, low-cut blouse, bangs and ringlets draped on her bare white shoulders. The pages are a testament to the belief Mother tried to instill in me: men are the leaders; women are beautiful and fragile; I abhorred her view and told her so.

But also inside the journal is a copy of her high school newspaper identifying her as a reporter and a list of the books she read in 1924 when she was fifteen. I've always kept such records and had no idea she did.

Did she tell me how many similarities we shared? When I was in grade school and high school I copied poems in my diaries and gave her poems I'd written. Surely she told me she had done the same. Did I forget? Or in the chaos of raising a daughter alone, had she forgotten her own childhood? Was I, like most of us, so obsessed with myself I ignored her? Perhaps my rejection of the romantic feminine idea simply disgusted or annoyed her so much she kept that part of her life a secret from me.

My mother was more like me than I ever imagined. No, I'm more like her than I had thought. What did I lose by not discovering these similarities until so long after her death?

April 19. At sunrise, the spring peeper frogs are loud and constant on the pond, as if they had arrived overnight. Two herons stand in the water, scratching their breasts with their beaks. Then they wade out into deeper water and wait, necks extended, for a frog. Thirteen cranes take off from the pasture as a quartet of geese come honking to land. The abundance of wildlife in this little scrap of prairie, hardly wild since our house, the cattle, a highway, and railroad occupy it, continually astonishes me.

I read today that an arachnologist from one of South Dakota's small universities discovered a new species of spider, the *Theridion pierre*, on the Fort Pierre National Grassland near the center of South Dakota. The discoverer, Brian Patrick, has found spiders whose location in South Dakota is the farthest west, east, or north they've ever been documented. That's because few scientists work on the sparsely-populated northern Great Plains, says Patrick. "It's not very sexy to work in South Dakota," he said. "I'm poor; I have to work in my back yard. Turns out my back yard is pretty fertile." The researcher said that in talking with colleagues, he's learned that, "I always have more diversity in my grasslands than they do in their forests."

The local newspaper gave this story prominent placement; what a great reminder that we should not be so hasty to encourage "developments" that include paving the prairie. This confirms, again, that we haven't begun to discover what rich knowledge the prairie might still conceal, exactly what I've been saying for years.

April 20. This afternoon we spent a half hour watching two buzzards who had landed on a south-facing roof in the ranch yard. They spread their wings and stood with their backs to the sun. Authorities say this behavior warms them and may also discourage parasites. Both then folded their wings and settled low on the roof.

Buzzards are easy to recognize in flight because the end feathers of each long wing are spread like fingers, and they can soar for a longer time without flapping than any other bird their size. I hope the buzzards will nest here; we even have a dead antelope nearby as temptation, though a golden eagle has been eating it.

Thinking again about the future of the ranch. I want it to belong to someone who will work hard to keep it. Yet, thinking of how hard we always worked, I wonder if I can find anyone for whom the rewards are enough.

Today the plum bushes are blooming just as they were, my journal says, on this day in 1981, but the joy of living in the country is only one small benefit of ranching.

Even inheriting a ranch isn't enough to keep young men and women here if their interests lie elsewhere. A popular myth holds that ranchers establish dynasties, filling vast acreages with family empires that remain strong for generations. In a few instances, this is true, but the case is sadly different for the majority of ranches in this neighborhood.

In my father's family, one son of three volunteered for war service and was killed. My father was educated and expected to go elsewhere; his older brother operated the ranch. My father worked in offices awhile, but returned and worked with his brother Harold until they each married and divided the land. Harold had no children. My father and mother hoped for a son but had only me. I have stepchildren, but none interested in ranching, so the era of this family on this ranch will end with me.

Other ranch women in the area could tell similar stories. One woman was, like me, her father's most reliable help while she was growing up. Like me, she loved the work, could ride anything, fix anything; she knew cattle and was always ready to help. When she told her father she deserved to have the land, he refused to give or sell her any part of it. Instead, he kept making excuses for his young son and eventually handed the place to him. The son bought a new pickup, married a woman who didn't care for ranching, and died in a drunken wreck. The widow sold the place. Meanwhile, the daughter had married a man who had never ranched but who was a hard worker. They bought five acres and a few calves; each year they bought a little more land, a few more cattle. The two keep scratching along on their little place, somehow hanging on, while her father's place has changed hands several times, going always to wealthier and more distant and uninvolved landowners.

A friend in Wyoming also told her father she wanted the ranch but her father said no, it would go to her brother. Her brother wasn't interested so the man sold it without consulting his daughter. She writes nostalgically about ranching, paints pictures of related subjects, and is involved with cowboy poetry. She has no children.

Let's see, another neighbor took over his parents' ranch and worked so hard he forgot to get married until he was in his fifties, and then selected a widow with three grown children who had never been in the country. After he died, she put the place up for sale. Her children persuaded her to auction it off either as a whole or in pieces, *whichever brought the most money.*

I wrote the story of the sale in the first chapter of my book *No Place Like Home*. The widow wandered around the room after the bidding bleating, "But I thought it would sell as a ranch!" That could not happen once the standard of "most money" was set. Most of the land is now subdivided.

I am not cheering myself with these reflections, nor am I making progress toward deciding what to do with this ranch.

April 21. We've found dozens of owl pellets under a juniper tree discarded by our resident great horned owls. At night and again just before sunrise we often hear both the male and female hooting in our windbreak. Since an owl can fly so silently, I wonder why it makes noise during prime hunting time. Is that a cry of triumph after eating a rabbit? Or a warning?

I've kept thinking about ranch inheritance. The Western myth insists that an ancestor came to the West, claimed (perhaps illegally) huge tracts of land, drove off the native inhabitants, hired itinerant cowboys to ride the range, hated farmers, and passed the ranch down to his descendants. According to legend these families became arrogant and rich, influencing legislators to increase their power. The reality here is different.

One of my neighbors, for example, comes from a ranching family in another part of the state and has several brothers. As they grew in age and ability, they traded work with other ranching neighbors, as is the usual good neighbor policy. Some neighbors appreciated the extra help so much they began paying the boys. Soon they were making money from the neighbors.

They received no salary at home, nor were they paid in cattle, as many ranchers, including my father, did. He advised me when to sell and trucked my cattle to the sale ring with his own at no cost to me. (But, I grumbled as I wrote a check for the property taxes a week ago, he never once showed me the property tax bill.)

When the boys questioned his policies, their father said, "I'll take care of you boys." They stayed home, worked hard, and a couple of them married. When they needed something, they told their dad and he bought it for them.

Finally the oldest son said to his brothers, "You know, I can do better working for the neighbors," and he took a job with another ranch.

Not long after that, the one I know best got married. He told me, "When you have to ask your dad if you can buy your wife a bed, there's something wrong."

So he found a series of jobs on ranches, enduring long hours and low pay for years until he came to this area.

My childless uncle Harold Hasselstrom lived a similar story. He had a reputation for never keeping hired men long at the time that his leg was amputated. By the time he got back on his feet, which he did remarkably quickly for a man in his seventies with one leg, he had learned something of being helpless. Perhaps he realized that his future depended on the kindness of others. He'd recently built a new house, so he moved his new hired man, Rick Fox, into the older ranch house on the place. When my aunt Josephine was diagnosed with brain cancer, Rick's wife nursed her much more faithfully than any of Jo's blood kin.

Harold spoke with his lawyer and called me, furious because he couldn't give the ranch to the hired man. "A man works all his life on a place he ought to be able to do what he damn pleases with it," he said.

So he sold it to the man he preferred for as little money as the government would allow, and the place is now the Fox Family Ranch. Not Hasselstroms, but a family.

My mother's Hey family ranch tells a different, and perhaps more typical story. My grandmother's first husband, father of my mother, was killed while she was young. Grandmother married a mechanic, Walt Hey, who took the little ranch as payment of a debt for car repair in the 1930s. When he died, my grandmother was left with three sons and my mother, who eventually took secretarial courses and left the area. One son, Bud, was a mechanic, and another, Harry, worked a variety of jobs. The third son, George, didn't particularly want to ranch, but when he came home from war, he began helping his mother, married a girl from the next valley and had a son and a daughter.

George gave his children advice I've heard often from ranchers, "You ought to get an education so you don't have to work as hard as I did." His son became an engineer and his daughter a musician married to another. Their children aren't interested in ranching, but the old folks are still there in their nineties. When they die, the place may be sold, though it's probably too remote to become a development for retirees; if it becomes someone's hobby ranch, the new owner will probably want a paved highway to make access easier.

The success stories of ranching are not family dynasties, but people who very much wanted to ranch and worked hard to achieve that dream—but often failed to convey it to the next generation.

Not all ranchers have been good land stewards, but those who are still in business after a hundred years have done better by their land than many of the newer, richer hobby ranchers can imagine.

More importantly, though, we need to educate the public on what it means for the nation to keep farming, ranching and food production as American businesses in this country. Raising animals in Confined Animal Feeding Operations (CAFOs), often with government subsidies, damages our farm and ranching economies as well as polluting our air and water.

Ranchers need to speak up to counter the ill-informed people who lump ranching with logging and mining and call it "raping the land." Probably young ranchers also need a better understanding of the urban world's thought processes and a broader understanding of different lifestyles, religions, and politics.

April 22. Earth Day. In perfect symbolism, I peeled the season's first tick out of my hair. The air and the grass ripple and quiver with little birds, zipping among the grass stems gobbling insects. A tree swallow zings past carrying feathers to its nest in the nest box on the electric pole.

On this day in 1981 Mother and I went to town where I read poetry to the Senior Citizens. When we got back, the electric poles were up for our house. We were settling into ranch life, expecting to spend the rest of our lives here, inheriting the ranch in the traditional way.

This morning Jerry and I drove along the Missile Road until we found the lek a neighbor had told me about; the grouse were dancing. These are not the big colorful sage grouse that make headlines in the southwest, but the smaller prairie chickens native to this area, and this is the first time I've ever been in the right place to watch them.

A half dozen males made booming noises, bobbing and weaving in a rough circle, chests puffed out. With binoculars we could see the yellow sacs on either side of their throats, puffed up with air. Now and then one would charge another and they'd collide into a tangle, tumbling and rising in the air, clawing at one another's chests. Now that we've found this lek, we'll come back every spring. One of my favorite Plains Indian dances shows how closely the natives observed their surroundings by imitating this mating ritual. Think how long generations of these birds have danced on this spot, though there's a highway nearby, a train track, and cattle.

April 23. The night sky is clear for the first time in days, the familiar constellations vividly silver.

Today at the post office, a woman was trying to rip open a package. When I whipped out my pocketknife and sliced it open for her, she looked shocked. I've carried a pocketknife since Mother explained to me what things women aren't "supposed" to do. I still regret that I wasn't allowed to take auto mechanics in high school, where I learned about transitive verbs, knowledge for which I've found no use. I've often had to work on my car, and I use my pocketknife daily.

April 24. 5:46 A.M. I love the soft blue-gray light before sunrise. The sky is deep blue to the west, lightening toward the east, where only a paler blue indicates where the sun will rise. Then the light blue slowly spreads overhead, like the tide coming in, until the east begins to show a line of pink.

Because I have lived and raised cattle here all my life, my body is composed of the dirt and grass and water that nourished the beef raised here. In Walmart today, watching shoppers waddle through the aisles in cheap, bright clothing, I decided that their bodies are made of plastic, Twinkies, and artificial colorings.

At 8 P.M., I'm leaning against my pillows in bed with a crossword puzzle, tired from a day spent partly in town. Being among so many people tires me more than any physical labor. Over my right shoulder a full moon brightens as the sky slowly goes dark.

Both dogs snuggle close enough to touch me, so they know their "pack" is together. Every night is a "Two Dog Night."

April 25. The dogs drop their noses and snuffle along vole highways, packed dirt paths two voles wide leading through the grass to pantries packed with seeds. With layers of snow on the tunnel roof, the mice are invisible to the winged hunters.

Today I picked the first rhubarb and made a pan of Rhubarb Dreams, a perfect celebration of spring. The pioneers relished rhubarb because it was one of the earliest and most reliable of garden produce after the long winter without anything green.

April 26. A pair of northern shoveler ducks float on the pond. Thirteen Sandhill cranes stalk the field. On our hillside, where cattle don't graze, the green grass grows up through older brown grass. Among the clumps of grass I see yellow globs. When I look more closely, I realize I'm seeing meadowlarks pausing in their eating to swell their breasts and sing.

My father's 1981 journal says that a man knocked on the door at 6 A.M. on this day because he'd had car trouble. My father didn't record in his journal what he didn't know: the rest of the story.

At that time George and I were living in a little addition on my parents' house. The knocking brought George upstairs from our basement bedroom, and I could hear the man saying something about his car breaking down.

Could he use our phone? Of course George said yes but I could tell he was uneasy. While the man called, George stood at the top of the basement stairs. Then we heard the man say, "I'm going to give myself up. They're going to catch me anyway."

George casually reached behind him and I, standing on the steps just below him, slipped his .357 into his hand. He tucked it under his shirt and kept his hand close to it while the man called a tow truck. Then he thanked us and went out the door. We watched him start walking back to the highway. We didn't call the police. We watched the newspaper for several days but saw no story about a law-breaker giving himself up.

April 27. Planted the rest of the potatoes; rather than digging, I laid them on the ground and covered them with three feet of old hay and straw. We seem to harvest at least as many this way as we ever did and don't have to dig them up in the fall. Last year, however, we created a Vole Picnic Ground; under the mulch, the voles dug tunnels to several potatoes and devoured them.

Reading my journals from the year we built our house reminds me how much writing I was doing then. I wrote an article about house-building while proofreading a book my Lame Johnny Press reprinted, *Buffalo Gap: A French Ranch in Dakota,* written by the Baron De Mandat-Grancey and originally published in 1887. I printed a limited edition of five hundred copies; the modern miracle of Google tells me #454 is available today for only $57.00. If I'd made that much money on each book the first time, I might have stayed in the publishing business.

When I read my journal and realize how much I was accomplishing in 1981 I am stunned: every day we worked on the house and I cooked meals, did laundry, helped my father with the cattle, and worked on the harvest—canning cucumbers and tomatoes, freezing squash, corn, tomato sauce, drying basil, parsley, and other herbs. Several times I "*canned 5 quarts of tomatoes while I fixed supper.*" One day I wrote, "*I am about through cutting up cucumbers and putting them in jars, a witless operation at best.*" I often wrote how tired and sore we were but we enjoyed many light moments: somewhere I have a photograph that symbolizes our year: a weary-eyed George, slumped in his filthy coveralls, sitting on the deck holding a gin and tonic.

I'd rather not admit what else I see in my own journals: that I was often edgy and bitchy and George was tense and grouchy. We were working hard on the house and were nervous about our future on the ranch as my father got less logical. We knew that once we finished the house we'd be out of savings and committed to remaining on the ranch. I might escape by traveling for workshops and talks but unless George found a job off the place, he could not.

My notes show me that we both understood how hard leaving the ranch would be and tried to be gentle with each other; I baked his favorite cookies; he made pancakes for breakfast. What would our lives have been like if we'd left the ranch? He'd have died anyway, and I'd have been established somewhere else.

April 28. Reading by the window, I heard the season's first mourning dove and glanced outside. On the deck railing sat two doves, hunkered low; feathers the color of early sunrise ruffled on their smooth round breasts. Romance is blooming.

Romance! Suddenly I'm scrambling through Mother's journals to find what she wrote in May 1925, when she was sixteen. Her mother has asked if she'd ever been kissed.

"*I said no,*" she wrote, "*and lied like everything but I don't care. She'd never understand that it is a modern custom, this goodnight Kiss stuff. She preached about B. might lose his control. Lord, whoever thought of it. B's the very soul of honor and Oh Hell she thinks she's smart about that losing control. They just do that in stories and probably in married life or something . . . this serious real life stuff makes me sick. All I want is romance. Kisses are romantic but Oh God help me to leave it there. I want Romance.*"

She often wrote about B: "*I love him because he means everything that is good and clean and big to me. There isn't anything little in his whole make up and when he comes home from college and if he still loves me and I do not love anyone else why we'll get married and live happily ever after. In the meantime I'll try to make all my littleness disappear and be as worthy of him as he is of me. I love him because he's all that ever was good in manhood and I give my heart to him forever.*" She did marry him a few years later. When she was in her seventies, she told me she'd never loved him. I found his letters after her death; he surely loved her.

Finally I can look beyond the mother who was disappointed in everything I did and see a woman who grew up in the 1920s, when "The Sheik of Araby" was the ideal man. Reading her list of the movies she watched and the poems she memorized, I sneered at their sentimentality, but I'm sorry now. Of course a girl raised in a dusty little South Dakota town might fall in love with those images and waltz into her life expecting Romance. And of course she'd be bitter after two unsuccessful tries at marriage when she found herself back in South Dakota peeling potatoes on a ranch for a man who might or might not be home in time for lunch.

She said she didn't love John Hasselstrom when she married him, but learned to love him; the evidence of the Valentine cards they exchanged indicates that is probably true. They sent sweetly romantic cards to each other all their lives. I know she had other suitors when we lived in town; most of them presented me with dolls when they came to call. She chose him because, she told me, her friends at the bank looked up his accounts and found he was a "rich rancher."

Thinking as a woman, I realize she couldn't leave one man unless she had another on the hook. After my father died she looked in the mirror one day and said, "Oh, I'm old," and stopped coloring her hair. When I moved her to the nursing home, several men were interested in her, but by then she'd given up.

April 29. The bloodroot is blooming. I planted peas and more radishes in the garden.

On this date in 1978, my father wrote in his journal, "*Linda gave us silverware for our Anniv. Thanks, Child.*" He must have believed I would read those words. He often seemed to keep the journal for his own reference, recording the details of managing the place, but surely he understood the value this knowledge would have to me if I ran the ranch. Perhaps I can believe that he forgot that plan only when his mind became unreliable.

By April the next year, 1979, George and I were married and my father's journal records that "*we discussed the ranch and if L and G are to be here.*" I was still writing as much as I could, doing housework and gardening, helping with the cattle. George was working full-time on the ranch. When my father decided to pay us $300 a month, exactly what I'd been getting before we were married, George said he would not accept working for free so we began talking about how and when to leave the ranch. Finally, George signed up for classes leading to a graduate degree so he could teach, but it was too late; the malignant tumor that killed him in 1988 had begun its work. For me, this was a huge and shocking new idea: I might not inherit the ranch.

We saw the first snake of the season today; on the Internet I discovered that what I have always called a "blue racer" is really an Eastern yellow-bellied racer. I've always known the little snakes are non-venomous, and that they eat crickets and moths, but I didn't know they eat rodents like voles as well as frogs, toads, lizards, and other snakes. Aha! Blue racers also eat birds; that explains something that happened sixty years ago.

I was walking in the cedar windbreak near my parents' house when I felt a tickle on my neck. I looked down inside my open-necked shirt to see a small blue snake lying against my belly, its head raised, tongue flicking.

Without much thought, I yanked my shirt loose and the snake slipped to the ground. When I described it to my father, he called it a blue racer. I've never forgotten the not-unpleasant sensation of that small body against my stomach. I didn't know until today that, if caught, racers bite hard, writhing, defecating, and releasing foul odors. I was lucky; the experience ended my fear of snakes. I respected rattlesnakes, but never was afraid again.

April 30. The wind died for a while in the night but this morning it's back, twenty degrees and whining, clattering, whooshing around the house, finding new spots to creak.

Cooking, I say to teach myself patience, is like writing a poem. With my first cup of coffee, I consider what the day will involve, and what to have for lunch. There's a slab of salmon and I found a recipe for baking it with garlic, basil and parsley. Potatoes: microwave first, then fry? Green beans left from yesterday. So one line of the lunch plan leads to another and another, just the way a poem develops. I check on Jerry's plans, and we organize ourselves.

I savor our routines, store them up against the day they change, as they inevitably will. What a privilege and blessing to be alive and mobile, to live in this house. I even enjoy the hard work of gardening, which I could not do without Jerry's tractor for tilling, and his creativity in creating our system of plastic pipes to water, not to mention his good cheer and his patience about helping me with all these jobs.

— May —

May 1. On May Eve, ancient Celts, and probably most of our ancestors, conducted ceremonies they believed would encourage the wheel of life to turn toward light and warmth. As if observing the holiday, a male red-winged blackbird alights on a tall mullein stalk outside my window. The stem sways; he spreads his wings for balance, red and gold wing patches glittering; he trills once to announce spring and flies away.

Inspired by memory, I rush to my office for one of my mother's photo albums and flip pages until I come to the group of high school girls wearing gossamer dresses, twining ribbons around a maypole. In the next picture all the girls stand in a line; most of them are smiling. At the center stands my mother. A long white veil is secured on top of her head with a crown of flowers, and she cradles a huge bouquet of red roses. "Queen of the May," she told me with a dreamy sigh when I first saw the picture.

Perhaps her life was all downhill from then; the Queen of the May became the mother of a gawky daughter who wanted to be a pistol-packing Dale Evans instead of the fairy princess with a wand.

Staring at the album, I'm overwhelmed with pity for her. She wanted romance but what she found was what most of us find: reality. She put her trust in love, but makeup and a mirror couldn't save her from disappointment.

May 2. Curly buffalo grass cushions my feet, looking dry and blonde as it has all winter, but on the south ridge, a subtle green shows.

My mother wrote on this day in 1956, *"A May basket was left on my doorstep tonight—a day late but the weather was more like May Day. Yellow and white flowers, willow branches and some ferny plants and grass, a woven handle from inner bark of cottonwood. Makes me realize spring is here."*

What a gift this memory is! I recalled she helped me make May baskets when we lived in town, showing me how to weave strips of paper into a cup because we were too poor to buy the commercial baskets my friends used. We'd fill them with small candies for my friends. I'd forgotten that I made them for her for a few years when we moved to the ranch, before our relationship soured. I must have made this one without her help, since it used natural materials instead of paper.

May 3. At almost 1 A.M., a ferocious west wind slammed against the house like a freight train dragging coal cars, clanking and rumbling for hours.

Lying awake, I naturally started thinking about my journal search, delving into my father and mother's journals. I'm looking at their lives from as close to the inside as I can now that they are dead. What am I searching for?

Do I expect to discover why they dissolved into shadows of the people who reared me? Am I looking for some scribbled line that will enable me to say, "Aha! So *that's* why they behaved that way and now that I know I can close these journals and get on with my life?"

By the time the wind and the real train hooted into the distance, I had no answer. So I went back to my mother's journals, looking for happy memories to wash away the sour tang of her angers and disappointments.

A clipping surely shows she tried to be cheerful each day:

> This day will bring some lovely thing,
> I say it over—each new dawn;
> Some gay adventurous thing to hold
> Against my heart when it is gone. –Grace Noll

Mother's leather-bound copy of *The Rubaiyat of Omar Khayyam* is small enough to fit in the palm of my hand, the covers worn thin. She marked in pencil certain lines in these fatalistic verses that I often heard her repeat:

> The Bird of Time has but
> a little way
> To flutter—and the Bird is
> on the Wing.

The Moving Finger writes;
and, having writ,
Moves on: nor all your Piety
nor Wit
Shall lure it back to cancel
half a Line,
Nor all your Tears wash out
a Word of it.

Another of her lovingly battered books, *Sonnets From the Portuguese* by Elizabeth Barrett Browning, contains a note from Mother: "Linda: this little volume was a gift from Paul to me, while we were in Houston." I was born in Houston, and there she climbed on the train that took us away from Paul, my biological father, after she discovered he'd been sleeping with a woman in an adjoining apartment. No pencil marks inside, only one rose petal from Mother's unwritten journal.

May 4. After such a wind, I always look out at sunrise, almost expecting the landscape to have changed. The chairs blew off the deck but such winds are nothing to the grassland. The precious soil is so thickly covered with deep native grass it's not going anywhere. As the sky grows light, I see a wall of white wind on the west, headed this way.

While my parents were writing in their journals, creating lives for themselves on paper, they were living their ordinary days, busy with exterior matters. Mother's daily round was cooking meals, washing dishes and laundry, cleaning house, shopping, and writing to friends, work she mostly did inside, away from the landscape and wildlife which absorbed me. When she left the ranch, she went inside again: to grocery and clothing stores, meetings with other women. When she moved to the nursing home, she said the main thing she wanted to do was read, that for all those years she'd wanted to spend more time reading.

I can understand that; I spend at least three hours a day reading even when I'm not reading work related to my writing.

Still, her journal reminds me that for many years she also did a huge amount of gardening. While I now concentrate on edible gardening, she planted flowers and trees. She asked my father to haul in truckloads of good dirt from our creek place and spread it in front of the house. There she tried to grow a lawn and planted flowers: mostly tough zinnias in every color imaginable. She wasn't studying the prairie, learning which native plants would flourish in a garden, while enjoying the natural beauty, as I do. She was trying to create a city yard: a major difference in our attitudes toward our surroundings. On the other hand, in the 1950s, finding information about native plants would have been much more difficult. Now that I think about it, most of the women on neighboring ranches tried to create a protected oasis of unnatural beauty around their houses.

Father built her a stone path I just rediscovered in the overgrown weeds. She planted delicate moss roses along it. I don't believe they lasted more than one summer. I've dug out the stones and attempted to level them to recreate the path.

For our garden spot, my father hooked my little John Deere 420 tractor to his ancient walking plow, then held the handles while I drove the tractor. With my help, Mother planted corn, tomatoes, peas, beans, spinach, squash, and all the usual country produce. She labored day after day in the fall to can and freeze it for our winter use. She transformed herself into the perfect ranch wife, but she didn't enjoy it. Now the walking plow is disintegrating beside the graves of my past Westies on our hillside.

While Mother worked inside, Father was really living on the prairie: driving, walking, riding horseback through the acres of grass that comprise the ranch. Every day he was studying the cattle as well as observing the wild animals, the pronghorn, deer, coyotes, porcupines. He knew this area intimately.

So did I, because I was with him most of the time when I was growing up. I hated being inside, doing the household chores. Now I've made my peace with most of them and actually enjoy some: I can think about what I'm writing while doing dishes. I learned to cook from Grandmother and from 4-H classes; Mother had no patience with my mistakes. Judging from the evidence of her cookbooks, which are another form of journal, she loved to try new recipes just as I do, though I think my father preferred traditional dishes and discouraged her.

Nothing made vacuuming attractive to either of us.

Today, I am missing the life of the plains while I am inside writing about my love of the prairie. I see a lot through the windows, but my sinews and lungs miss riding a horse or walking through it every day.

Was it Thoreau who said that it would take seventy years for anyone to know a place twenty miles square? I spent some seriously formative years here, *knowing* the landscape in the most intimate ways possible. While I was gone for almost two decades, I continued to write about the prairie.

In the years since I moved back, I've been so intent on my work of writing about it that I have been unable to be in and of it as much. Always this tradeoff: if I were to walk the pastures more often as I once did, I'd enjoy life more and write less. Yet if I wrote less perhaps I would fail in my self-appointed job of helping to preserve the prairie life.

May 5. Grasslands are a symphony, and buffalo grass is the bass, the deep-throated foundation of the life that has evolved here. Of all the plants on earth, grasses are the most important to humans. All the plants from which we make bread—corn, wheat, oats, rye, barley—are grass, as is bamboo, as well as the bluegrass of lawns and the timothy and redtop of pastures. Yet we treat grasses as though they are background or wallpaper, not the center of our survival.

Buffalo grass, *Buchloe dactyloides,* is only a few inches tall and called "gray-green" in the reference books, but looks gold to me. As a dominant species of the shortgrass prairie, it spreads to form dense mats, holding soil from blowing or washing away with roots that may reach five feet deep. Named because of its value to the bison that roamed the prairies, it benefits from heavy grazing and trampling, though we're careful not to overgraze it. Native prairie recovers from drought more quickly than introduced grasses.

May 6. This afternoon I was sitting in a chair by the greenhouse, enjoying the feel of the sun on my face at almost sixty degrees, when I heard a high-pitched whine in my right ear. Is it the onset of deafness, I wondered? No. The season's first mosquito. I'm not ready!

"I tell you that the great cities rest upon these broad and fertile prairies. Burn down your cities," said William Jennings Bryan on July 9, 1896, "and leave our farms and the cities will spring up again as if by magic; but destroy our farms and the grass will be in the streets of every city in the country." Bryan's "Cross of Gold" speech is called the most famous speech ever given, but it's usually quoted about the gold standard he supported. I think the statement about farms and cities is considerably more important and ominously forecasts our current lopsided situation.

Several recent articles suggest that the population shift from rural to urban is having serious effects on the nation's priorities. Even the Agriculture Secretary says rural Americans are less relevant in the country's increasingly urban landscape. At least fifty percent of rural counties have suffered population decline, and more than eighty percent of lawmakers don't represent rural areas, so those who do represent us have a hard time convincing urban counterparts our needs are important. Too much of our food now comes from places whose citizens may not, at the very least, have our best interests at heart.

May 7. Thinking so much about my mother as I read her journals. Unfortunately, one of her greatest influences on me was to make me forever insecure about my appearance. Even for quick trips to get the mail, I feel guilty if I'm wearing work clothes.

The saddest part is that she might have taught me something about rising above disappointment and how she managed life as a single mother. She was so critical that her legacy is negative. I have to work at assuring myself that nobody is looking at me very hard and that, anyway, I look fine. Which reminds me of another friend's saying: "It'll be fine. And even if it's not fine, it'll be fine."

On this date in 1988, I learned that my father had prostate cancer and, symbolically perhaps, while we were giving scours medicine to calves, one kicked me in the head. I was exhausted, barely getting through each day, pushing to try to write at night or in the early morning on my South Dakota history book. Meanwhile, George was going to physical therapy, and we didn't yet know that he would not live out the year. In June, I learned that my first book, *Windbreak,* had been the subject of what the court later required me to call "copyright infringement." A romance novelist copied more than 75 passages from it into her published book. My 45th year was not looking good.

May 8. On May 8, 1955, when I was twelve, my mother reported in her journal that I had been up late the night before but got up early and made her a crown of willow, lilacs, and apple blossoms for Mother's Day.

My Mother's Day has been pleasant but unconventional; I didn't hear from any of my four stepchildren. I did some laundry, made a huge pot of green chile, read, and wrote; planted nasturtium seeds among the radishes and lettuce in the herb garden, and moved some sedum into the rock garden. I'm putting other keepsakes there too, things that clutter the house: statuettes, tarnished silverware, beads I meant to make into a necklace, driftwood, rusty bolts, keys to unknown locks. Sand dollars and other shells I picked up on beaches in Manzanita, in Maine, and in Scotland. A spoon George carved from bone rests in a bowl Jerry hollowed out of a pine burl. All bring back good memories that flow through the warm spring air like the songs of the blackbirds and meadowlarks. I won't have to dust or store these things, and by the time I've forgotten their meaning, perhaps they will have dissolved back into the earth, a disappearing journal.

Everywhere I've gone during the past few days, I have wished a "Happy Mother's Day" to every woman older than I am that I met. Several of them sounded surprised as they said "Thank you!" Today, staying home and mothering my garden and yard, I met no one.

I've been remembering my mother and the only grandmother I ever knew, Cora Belle (Pearcy, Baker) Hey and women like my aunt Anne who taught me as much as any mother could have. Some years I've heard from my stepchildren, but I shouldn't expect to since I don't write them regularly.

I've been thinking, too, about women who, for one reason or another, are not mothers. The reasons vary and I don't suggest that we declare a Stepmother's Day, or an Unsuccessful Mother's Day, but I like to remember that women who didn't give birth have made other contributions to the world.

On this day in 1988, already a difficult year, I had put a little wood in the firebox of the wood furnace and sat down with a cup of coffee when I heard a roar like an airplane landing on the roof. I looked out the windows and saw smoke and cinders rolling down over the deck: a chimney fire. I did nothing right: first I rushed outside to see if the roof was on fire, then called the fire department, then loaded computer disks and photograph albums into the car. I should have been putting water on the roof and covering the chimney. The first volunteer fireman who arrived climbed the ladder I'd leaned against the house and stuffed his Sunday jacket in the chimney, smothering the chimney blaze that might have destroyed our house.

We were still waiting to hear from the doctors the details of George's malignant tumor, and we had learned that my father's heart had a malfunctioning valve.

As I sit in bed writing in my journal just before sliding down to sleep, I hear the winnowing snipe calling gently in the cool air. On the slope below the house, a cow has just calved: I hear her murmuring to her baby as she licks him clean.

May 9. Forecasters said we are under a winter storm watch today. Early this morning the clouds tumbled across the sky, rolling gray and white like breakers on a beach, but I could see a hint of blue above them. Now the hills to the west are blurry with drifting snow. A few birds flash past the windows, but most must be in shelter. Last week, hedging my bet, I put away our winter coats, but left the snow shovels by the door.

All morning, as I work in the basement, the snow falls, heavy and wet. At first it melts into the dry ground, but by noon it is sticking, piling on the fences and electric wires. When I walk the dogs they get soaked to the skin and come in shivering so I have to put them in a sink full of warm water. By afternoon I can measure six inches of snow on the deck, and the wind is drifting it deeper on the lee side of hills, like the driveway. The temperature drops into the thirties. I pick a newly-bloomed wild iris, but it shrivels.

May 10. No birdsong this morning; the window screens are plastered with heavy snow, drifting with the wind. Long ridges of snow reach across the yard. The dogs flounder and then turn back. I try to shovel some snow off the deck—Jerry is away—but realize I would hurt my back too badly for the job to be worthwhile. I start out walking along our road to see how the drift is on the lee side and am soon up to my knees; my car couldn't possibly break through the drift. So I work inside and throw the ball for the dogs every few hours.

Late in the afternoon, when the sun is out, I see Rick drive through the ranch yard. Grateful for my cell phone, I call and ask him to drive to my house to break the trail. He gets stuck several times, but batters through. We visit in the yard a while before he leaves, and gets stuck again. He has no shovel, but he keeps roaring into the drifts until he's free. Even though my car has 4-wheel drive, I'll wait for more melting before I try.

May 11. I woke at three A.M. and saw an ambulance silently racing north, red and white lights reflecting off low clouds. I appreciate the consideration of a driver who reasons that the siren isn't needed at that hour on this deserted highway, especially since people out here must get up soon.

My mother's journal for this date in 1973 says, *"Linda called to say she is divorcing D—. Best news we've had in six years."*

Despite her happiness about my divorce from my first husband, she pronounced me damaged goods and said I might as well move home and help my father because no one else would ever marry me. The statistics supported her viewpoint.

May 12. Woke up with "And I don't know who I am but life is for learning," running through my head. Laugh, but those old songs sometimes have a kernel of perfect truth. I hummed the tune until I had to look up the lyrics. Yes:

> "Got to get back to the land, set my soul free.
> We are stardust, we are golden, we are billion-year-old carbon,
> And we got to get ourselves back to the garden."

Getting back to the land sure worked for me, all these years after Woodstock.

By noon today, most of my trail to the highway was nearly dry, but each low spot was a pond: two ducks swam in the deepest one and two more in the roadside ditch.

May 13. My 1988 journal for this date says, *"Our method of ranching involves so much attention given to individual cows, knowing their health, peculiarities, heritage. That's not the 'modern' way."* Some "producers," as ranchers are apparently called these days, regard cows as beef or milk factories. Our personal attention kept the cows healthy, provided more wholesome meat, and maintained the land better than modern "agri-business," the industrial-strength "operations" (they aren't really ranches) that operate with hired hands who have no ownership of the land.

My father usually checked every cow every day, visiting every pasture no matter how long it took. Ranchers who believe they can't make a living without thousands of acres and hundreds of cows may think they don't have time for individual attention to each cow, but many of them have time for dances, drinking coffee in town, and going to meetings to discuss how hard ranching is.

My father's methods make me think of Thomas Jefferson who envisioned a country of self-sufficient farmers and ranchers. Perhaps we should talk about "slow ranching" as well as "slow food." My father kept around two hundred cows on about three thousand acres, which included both grazing and hay ground; this was ample work, but not too much, to sustain one family comfortably. Our modern standard of living, though, seems to require enough "disposable income"—what a strange term!—so each member of the family can buy a new car every year and own a "smart" phone.

My other exemplar for proper land care is Wendell Berry. My father practiced, and taught me, "husbandry" as Wendell Berry defines it: using with care, keeping, saving, and conserving. Everything fits together; we cannot conserve our ranches, cattle, and homes unless we also conserve wildlife. The Lakota say, "*Mitakuye oyasin:* We are all related."

May 14. Today when I walked into the kitchen of the retreat house, a dishrag hung over the faucet just as my mother used to leave it. For a dizzy moment, I thought she might be in another room. The kitchen still contains her cupboards, the yellow countertops she picked out sixty years ago. Are these hallucinations—I've had several—caused by reading her journals? Perhaps I wish I could talk to her as an adult, without the confusion that always lay between us.

I'm still reading Wendell Berry and finding words relevant to my thoughts on grasslands. "To countenance mountain removal in Kentucky and West Virginia," he says, "is to agree to the eventual destruction of Yosemite and Yellowstone and the Smokies." He adds that we can't preserve the fine arts if we destroy "the domestic arts of farming, forestry, cooking, clothing, building, home-making, community life and local economy." We can't save historic buildings, he says, "outside of the context of the community and economy for which they were built."

I think Wendell Berry would approve my reading of the journals of the people who have preserved and conserved this place, as well as my remembering how we used to do things. Some changes in methods are improvements and should be implemented. But we need to evaluate what effect a change will have on the whole of the place before adopting it. Everyone chuckles when I mention the work we did with horses, but bringing the tractors to ranches shifted our economy from self-supporting solar energy to dependence on foreign fossil fuel. We raised the horses and fed them on our own grass; the tractor was built elsewhere by paid labor and required cash payments for fuel from another place. Understanding the implications of these changes is part of the job of the elders in advising youth.

"I now suspect," writes Berry, "that if we work with machines the world will seem to us to be a machine, but if we work with living creatures the world will appear to us as a living creature." Surely if this applies to ranchers and farmers, it applies also to people who spend hours with the personal machinery of phones and computers and TVs.

May 15. At sunset, a few drops of rain tap the porch railings. Blooming flax makes pools of blue in the grass. Walking the dogs, I spot a curlew nest under a limestone outcrop: two mottled eggs in a limestone bowl that is tilted to catch sun when the bird is away. Humans spend a lot of money understanding solar power; perhaps animals know it instinctively.

We sit on the deck with drinks watching the setting sun light up the tall cured grass on the hillside, hoping to see an antelope or two. Ducks float on the mirror of the pond. A snipe call shivers the air.

I tell Jerry what a hard time I'm having writing about my parents and their journals. He nods; what else can he do? He's a wise man, to know I'm not asking for help. I've read somewhere that one of the big communication gaps between men and women is that men want to *solve* things when women just want them to *listen.*

Watching as the ducks bob among their reflections, I realize I don't have to write another word. I could live here and enjoy this prairie without making any further efforts, possibly doomed anyway, to save it or create appreciation for it in people who have not had this experience. I have worked hard enough all my life; perhaps I should just *enjoy* the plains for the rest of my life without feeling guilty.

May 16. On this date in 1981, I drove to Newcastle to present a workshop; a terrible drive in the middle of a spring blizzard. The visibility was about six feet in front of the car as I drove through Custer State Park, and every now and then a buffalo loomed up in the road like a boulder. I managed not to swerve. If that storm were raging today, I'd call the folks in Newcastle and explain why I was unable to come.

The next day my father was furious because I wasn't there to help him with two sick calves. I'd told him I would be away teaching, but this was a portent of the way things began to go wrong between us on the ranch. I was editing an environmental newspaper, teaching writing workshops for cash, and running Lame Johnny Press as well as trying to sell my self-published book about self-publishing called *The Book Book.* George was working hard at ranching and feeling unappreciated.

At 10 P.M. in this present time, the chorus frogs are trilling loudly in the dam below the house, drowning out the truck traffic. We haven't seen fireflies around the dam for several years. Authorities suggest human traffic, light pollution, pesticides, and development may drive them away. I miss their flashing mating signals.

May 17. My dream last night was the first pleasant one I've had about my mother, perhaps ever. In it I was helping her clear out a room where she'd been living in the nursing home, and we were looking through dozens of gorgeous quilts she'd bought at the bazaars the church held to raise money. We held up each quilt and exclaimed over the intricate stitching, smiling at each other, and actually enjoying each other's company.

I woke up chuckling, remembering the time my mother was bidding energetically on a quilt and people began to whisper and smile and finally to laugh.

My father elbowed her, but as she grew angrier at the person who kept raising her bid, she refused to look at him. She kept shooting her hand in the air. The bidding reached a hundred, two hundred dollars.

Finally the auctioneer said, "Do you think this is a good time to tell her she's bidding against herself? Have we raised enough money on this one?" The crowd guffawed and Mother was furious. If my father hadn't held her arm while he smiled at the auctioneer, she'd have run from the room.

When I woke, I could see several of those quilts, hanging on the rack Jerry built. Several others are in the retreat house where quilters who come to write tell me I shouldn't be using them or they will wear out.

Surely the dream is a result of my reading and writing about her journals lately. Perhaps the reconciliation, even in my sleep, will allow me to put aside the anger between us.

May 18. Today in my mother's journals I noticed how often she wrote happily of something I'd done. May 22, 1957: *"Linda got breakfast this morning and I went back to sleep. When I got up she got my breakfast, had a chicken picked. Yesterday she helped John butcher and made a blouse."*

On September 4, she noted, *"SD champion oral speller and all time champion of my heart started Hi school today, a thrilling experience for her and for her parents. We were 'bowled over' when we heard it over news last eve that she had won spelling."* I remember that the contest was held in Pierre, and I must have gone with a teacher, but I apparently didn't call home with the news.

I was a busy little overachiever. I submitted an essay to *Family Weekly* and won third place in the county contest for my essay on the subject "Why Alcohol is a Menace to Highway Safety in South Dakota."

In early 1958, Mother remarked, *"Seems strange to have my daughter wearing my clothes and so well,"* and in June she noted that she bought me my first high heels, baby dolls. In December of 1958, when I was fifteen, they went to town shopping and *"left Linda alone, making candy and cookies. First time we've left her alone but home soon after noon."* They trusted me, but not too long.

On January 5, 1959, she wrote, *"I was awakened by John this morning asking if Linda was in with me. He'd been lying down in living room with his cough. We looked all over but no Linda. He started out back door and called; there she was out chopping ice off truck—what a girl!"*

A few days later she noted, *"Linda is famous! [Arthur] Godfrey read about her letter to J. L. Wallington of Ketchum, Idaho about Badger Clark poem 'Boastful Bill.'"* I recall this incident only dimly; I believe Wallington had asked Godfrey, who had a popular radio program, about the authorship of a few lines of verse. When I recognized them as the Badger Clark poem "The Legend of Boastful Bill" (I still recite it while driving) I wrote to Wallington, and he told Godfrey, who mentioned it on his program.

All these entries of Mother's seem so cheerful, as if she was completely content. Why is it, then, that her handwriting begins to get smaller and smaller in her journals throughout 1959, when I turned sixteen?

By early 1960, she was writing two lines of script between each pair of ruled lines. Her note just before Christmas may be a partial explanation: apparently I was becoming an unruly teenager: *"Terrible row with Linda because I wanted her to wear her red dress and be 'dressed up' a little this last day of school before vacation. Makes me wish I could toss out the tree and take all her presents back to store. So happy last night when I kissed her soft cheek after she'd been asleep and lost all her hostilities."*

She was still so angry the next day that she screamed at my father to throw the Christmas tree and presents away, but he talked her out of it. This over-reaction was typical anytime we disagreed. My part of the argument probably involved me sticking out my lower lip—she hated that—and her screaming. I'd learned long before John Hasselstrom came into our lives that to argue with her earned me a slap on the face. If she knocked my glasses off, she slapped me again. So I usually confined my unhappiness to sulking, to rolling my eyes, to mumbled answers, to that hanging lower lip and scribbling in my own journal. All these actions drove her to more and more screaming.

May 19. Flipping through more of Mother's journals yesterday, I watched her handwriting get smaller. My father and I knew she wrote in her journal sitting on the couch each night, but we had no idea what she was writing. What was she feeling?

Surely the tiny handwriting was a blatant symbol that she felt small, unappreciated, perhaps depressed. Outwardly, she just grew angrier. I can begin to trace her mental deterioration, too, by the increasing illegibility of her handwriting and her reminders to herself of things she once knew. Snippets of information she pasted into the journals provide insights: pithy sayings, dozens of obituaries (most frustratingly undated), photographs of her favorite male politicians and movie stars.

Could we—my father or I—have prevented her decline if we'd paid more attention to what was important to her?

May 20. Mother's journals become too painful to read in 1973, the year she seemed to lose touch with reality for a long time. On Mother's Day she wrote, *"Mother's Day card from my daughter. I hear from her so seldom, forget she's only a few miles up the line.* [I was living in Spearfish, about 65 miles away] *John called her yesterday to tell her I am sick. If it was* John *she'd have been here last nite. I raised her to help him tho and glad she loves him as a father, I think. Or is it just his money? I don't know her anymore."*

If my mother believed I loved my father only for his money, either she was completely deranged or she didn't know me at all.

Along with my father, I was slow to grasp how angry and mentally confused Mother was that spring and summer. He probably didn't want to believe what was happening; my excuse is my own situation. After six months on the ranch, my first husband wasn't doing ranch work and couldn't pay his child support, so I got a one-year teaching job and we moved to Spearfish, South Dakota. He started looking for work, he said, while I paid his child support and our daily expenses.

One day while I was still denying my suspicions, I reached into the pocket of his car's passenger door for a map and found lacy underwear. Someone told my mother about seeing him in public with other women, which didn't help her mental crisis. I started divorce proceedings and concentrated on finishing my teaching year, but drove home every weekend; Mother was never satisfied.

Now I wonder if she was reliving the pain of her earlier marriage to my biological father. Her journals are filled with red ink: *"Just being self, which is bitchy."* We tried to talk her into going to the hospital and several times I met them in Rapid City to go to dinner at my father's urging. Mother suspected us of plotting. In erratic scribbles she wrote that if the car turned one way she'd know we were going to the hospital.

"If you turn toward the hospital," she told us once as we headed for town, "I'll start screaming, and you'll have to get a strait jacket."

She decided that I had brought cockroaches into her house, so she closed all the doors and windows and sprayed an entire can of Raid. *"The joint has been Raided,"* she wrote in her diary. My father was beside himself with worry and fear.

One day she started shrieking that we were trying to kill her. "I want to see my mother just once more," she screamed, so my father took her to the ranch where her mother and brother lived, left her there, and went to Rapid City to try to get a doctor to admit her to a hospital. No doctor would see her.

When she found out what he'd done, she wrote, *"I raised hell and said you can't trust any man and John turned on his heel and said he didn't have to take that."*

Apparently I talked her into going to the hospital, and as soon as we got there she started calling relatives to get her out. I don't recall and didn't record in my journal the details but she was only in the hospital a few days. Throughout that year she writes about taking various medications or refusing to take them and about taking sleeping pills every night. My father took her to visit doctors and psychiatrists, sometimes cursing them, sometimes confused. If I came to visit, she spent our time together screaming about my neglect or telling me that if I had a baby, she'd be happy to raise it.

My own journals of that time don't help me recall much, filled as they are with anger at my husband, fear of being alone, and general confusion.

"Why," I wrote once, *"when I come to the ranch, do I always open drawers and trunks and cupboards looking for some lost soul in a corner, or the key to my life."*

Was I looking for a relationship with my mother? The comfort I expected from marriage? I found neither.

May 21. Blooming: Johnny jump-ups, bluebells, gumbo lilies, and phlox.

On Saturday, May 21, 1955, when I was twelve, my mother wrote, *"First day of vacation. Linda just a little girl in spite of some maturity."* All her life she thought of me as a little girl; perhaps that's natural for mothers.

Mother was never normal after her hospital stay, and everyone was afraid to disagree with her for fear of launching her on another round of screaming hysteria. Relatives hosted family dinners on holidays. My father took her "out to breakfast" several times a week; on nice afternoons, they played golf somewhere. She was Queen of the May again, an utter tyrant. If she was displeased, her voice would rise to a screech and she'd begin to shake. If that failed to move us in the direction she preferred, she'd shift in seconds into shrieking and throwing things. My father couldn't escape. My efforts to help her were useless. First she said that I should stop teaching and come help my father since no man would want me, then that I should have stayed married to my loathsome first husband because she wanted grandchildren.

During that awful summer of 1973, I also met George, who would become my second husband, "the good one," as my friends said. How in the world did George have the patience to stay with me through the maelstrom of that time? I cannot recall or learn from my journals how I survived.

May 22. I realize Mother behaved like a spoiled child most of her life, but she didn't learn that from my grandmother. My childhood response was the opposite; I refused to cry when she hit me. When she vowed "to make a lady out of me," I became a "tomboy." My reaction to her behavior was to keep my worries to myself and my journal, showing as little emotion as possible to the rest of the world. While I don't believe it's necessary to emote in all directions, I do think it's possible to keep so much pain bottled up inside that one creates even more suffering. Some authorities suggest that locking our injuries away causes physical illness. So my journals, as chaotic as they sometimes are, may have kept me more sane and healthy than I might otherwise be.

May 23. My journal for 1982 tells me that George and I were moving furniture and finishing up details in our new house. My parents went to Edgemont to meet Mother's brothers at Grandmother's grave, but we stayed home. I wrote in my journal that since I visit with Grandmother every day while I use her wooden spoons and bowls, I don't need to visit the cemetery. We decided at noon to sleep in the house for the first time that night and continued hauling furniture through a hailstorm and rain.

"So much space!" I wrote, but we had no kitchen shelves. Our first supper in the dining room was leftover stew. The stereo played; what, I wonder? The hillside glowed vivid green and gold at sunset; frogs and killdeer sang around the pond.

Reality struck the next day; we got the results of a test on our well and discovered it contained coliform bacteria. We pumped and dumped bleach down it until it was pronounced safe. Since the well is high above the corrals, and had been fine for years, bacteria must have gotten in when we were hanging the submersible pump.

I try not to think of that inauspicious beginning as I run a glass of cold water from the faucet. In town, folks depend on authorities to keep their water safe, but safeguarding our water is one more job for country-dwellers. I wonder how many of the new country folks understand this?

May 24. I wake up tense from the hum of traffic all night. Another deer lies dead on the highway beside the fox killed yesterday. The state condemned good pasture to pave and widen this road so that people who refused to obey speed limits could speed more conveniently and safely. In exchange, the prairie gets disruption and death. Buzzards cruise above the borrow pits, a sure source of dead animals left by humans in a hurry.

In the evening, I was reading in the living room while Jerry took a bath. I got up to get a drink of water from the kitchen. Something leapt from the kitchen into the pantry.

Mouse, I thought.

No, it was a toad, nearly as big as my hand. I picked him up and took him out to the herb garden and then started wondering. We've had a toad in the raised beds near the basement door, but he seldom ventures out of them. If it's the same toad, in order to get to the kitchen, he had to get inside, hop across the basement and then hop up sixteen steps, each nine inches high. The basement door has a device that slows its closing, so it's possible he got inside when I turned away from the door as it closed. If he came in by the east door of the house, he had to hop up several steps to the deck, and then inside and across the living room where I've been for more than an hour.

I've seen Cosmo carry a baby robin so gently in his mouth that it wasn't harmed, but would he carry a toad? Wouldn't it taste bad? Surely both dogs would have barked if they'd seen it. A mystery.

May 25. Today I sorted letters in which my mother told me more about why she divorced my biological father. Mother and Paul had separated, and grandmother Cora was staying with us when Paul's sister Ruth came to visit. One day, Mother left me with Grandmother and went shopping with Ruth. When they stopped for a beer, Ruth started crying and told Mother that Paul planned to kidnap me and raise me so that I could support him.

Ruth admitted that she'd planned to help him until my father said, "My daughter isn't going to get in trouble. I'll teach her how to have intercourse without getting pregnant."

One day Paul borrowed Mother's car and wrecked it. He confessed that he hadn't been alone, but not which woman had been with him. Next day Mother saw Thelma, her upstairs neighbor, hanging a bloody white dress on the clothesline they shared in the back yard, and realized the truth.

While she was waiting for the divorce, Mother moved to Edinburg, perhaps to be close to her mother's sister. When Paul came to visit, he took me shopping on foot because he'd lost his driver's license. I couldn't have been more than three, so we probably didn't walk far. He bought me a doll; I have a photograph of me having a tea party with her. When we got home, Mother wrote, he was so drunk he flopped on my aunt's big bed and slept until he was sober, then caught the bus back to Houston.

I've re-read these details in Mother's letters, looked at my own journals where I recorded all this once and now I've written it all again, and my primary emotion is embarrassment, and amazement that I remember none of this. This was no soap opera or movie; it was my life. Certain details I do remember perhaps confirm what I don't recall: I crouched under a kitchen table while my mother screamed at my father and smashed liquor bottles in the sink.

How many other memories might be in lurking in some curdled corner of my brain, affecting the way I have viewed men or marriage or some other aspect of life ever since?

What do these memories mean to me? My mother was a naïve young woman raised in a normal household. She educated herself; she went out into the world with confidence, and she probably didn't do anything really evil, yet these things happened to her. I can't condemn her for anything but perhaps loving unwisely, hardly blameworthy since it's pretty common. She told me that I was "damaged goods" after my first marriage, so surely she felt she deserved my biological father's treatment of her; she wasn't good enough for anyone else.

What do I do with this knowledge? Remember, surely, and be wary of times that I might be reacting to something in my past.

And how did it affect my mother? Did she blame herself for not being able to rescue the man she loved? I will never know now. I have all the information from her I am going to get; the only journals, letters, and memories left. Nothing new will come to light.

May 26. A few branches of the lilac bushes in front of the retreat house are finally showing blooms, probably damaged by the snowstorm. I remember them as walls of lavender, a poem to spring, planted in the hard ground as a symbol of my father's love. I don't take care of them as well as he did, so they may be dying.

Still wrapped in thoughts of my mother, my father, and my first husband, I worked on a poem I started several years ago. I know some poets who write a poem and publish it within a month or two. Surely it hasn't gone through enough thought by then, or am I just slow to make decisions?

May 27. If not for my journals, I wouldn't have saved some of the memories that are helping me create healing poems now. Writing details, writing every single day, helps the brain to record and keep memories that one can take out later, examine, perhaps put to rest or use creatively.

When I think of the damage my mother did to me with her hysterics, I immediately think of the worse damage she did herself, destroying her own happiness, making my father miserable. Worse, the pattern she followed recorded itself in my brain, so sometimes I can feel myself slipping into the kind of outburst on which she thrived. I've heard wild accusations like hers coming out of my own mouth too many times. By writing about her, by seeing what she did to herself and others, I've begun to be able to stop myself before I launch into the destructive performance she employed to get her way. I've explained the pattern to Jerry and he has helped me.

May 28. Sunrise: a doe antelope tiptoes up out of the draw below the house, followed by her fawn, so new it's damp and staggering. Each time she stops, it fastens its mouth around a nipple and gets a little nourishment while she looks at the house as if it wasn't there last night when she began giving birth.

On this day in 1982 the nighthawks came back; *"we hear them tonight around our high house,"* I wrote. The birds soar high before they drop, booming, to snatch millers and mosquitoes. Experts differ about whether it is their wings or their beaks that make the distinctive sound.

I didn't appreciate my mother when I was growing up because her skills were not those of a rancher; only now do I begin to see how unhappy she may have been. I tried not to do any of the things she did that seemed selfish and self-centered to me and yet sometimes I catch a glimpse of myself in a mirror and see her. We can't escape our ancestors even if we want to. Or maybe especially if we want to.

Thinking of my mother has made me think of other women in my life from whom I learned how to survive. My aunt Josephine had no children and insisted I was "not a real Hasselstrom." A hard-riding, cussing woman who owned her own cattle and sheep, she kept her own checkbook and worked alongside Harold outside. When they came in at noon, he rested while she fixed the meal. Yet she encouraged me to do what I wanted to do in exact opposition to my mother's hopes.

When I was married for the second time and kept my own name, Jo insisted on calling me "Mrs. Snell." She was a feminist in every conceivable definition that's important, but she wasn't going to call herself one, and she wasn't going to call a married woman by her maiden name. She was so tough we had to remind her to die when she had brain cancer.

My grandmother Cora never stopped listening to me, encouraging me, and giving me good advice. Though none of those women owned ranches except by being widowed, they all inadvertently prepared me for the job of owning this land.

I think the view of women as incapable of owning land was similar to the view that women's writing wasn't worth reading, especially if it was about their daily lives. When a woman like my mother wrote about her ideas and work in her journal, she was actively claiming the right to be heard, even if she didn't realize it. She sensed the importance of recording details of her life for an unknown audience. She never told me to destroy her journals; she meant them to stand as some kind of record of her life.

Those other women who taught me how to live mostly recorded their lives and loves in other ways: in quilts, in jars of vegetables in the cellar or freezer, in embroidery and gardening and taking care of other people in their communities. Their work was their art, and their journals; their labor was the books they didn't write.

May 29. A red-tailed hawk shoots up out of the cottonwoods, pursued by two red-winged blackbirds; the small birds dive from above and below, pecking, swooping as the hawk screeches. When they reach the edge of their territory, another pair rises to take over the chase. The hawk flies awkwardly, carrying the long body of a writhing snake.

My father wrote in his journal in 1989, *"Our anniversary* [their thirty-seventh]; *we are asking for a few more years together."*

That fall, I began teaching at a nearby college, sticking to the plan George and I had devised together of getting work outside the ranch because my father was paying us so little. Each day that I went to work, my father wrote in his journal that I had *"gone to town;"* he didn't seem to understand that I had taken a job or why I had done so. *"She's spent every day this week in town,"* he wrote, detailing the jobs I hadn't helped him with. What in the world did he think I was doing? Apparently Mother didn't understand either.

After Mother's bout of mental confusion, my father did take more time off. I believe that was part of her complaint: that he worked too hard and too long. She'd endured years of dried-out dinners and worry when he wasn't back from the pasture by noon; of course they had no cell phones then. Perhaps she believed that keeping George and me here to work the ranch would keep both of them safe from their declining health. That's how many rural families operate as the elders decline: they can live in their own homes because "the kids" take care of them. One rancher I heard lamenting "the boy's" mistakes was ninety and still the boss; "the boy" was sixty-five.

On May 30, 1953, when I was ten years old, I wrote in my diary about our annual visit to the cemetery to weed the graves and place flowers. My father would wander around, pointing out the graves of his mother and father, his mother's first husband, his half-brothers.

"I'm just visiting my future," he'd say. Standing over the graves of some former neighbors, he said once, "They used to be rich. They used to be somebody."

This was his way of introducing me to relatives I would never meet. After years of visits to the cemetery, I can now match some faces in our albums with the names on the headstones.

Now, just as I am responsible for what will happen to this ranch, all those graves are my responsibility. A couple of years ago Jerry and I noticed that a lilac bush on the graves of the grandparents and uncles I never knew had overgrown not only their graves but neighboring ones as well. We spent a day chopping, hauling, sweating, and mumbling to ourselves, clearing our family graves as well as the neighboring ones. Those turned out to be the grandparents of our neighbor just north of the ranch, so it was good to feel we'd done a neighborly deed.

Each year, I weed the graves, place flowers, and straighten stones. The plot my father bought when George died has room for both me and Jerry, though he has talked of being cremated. Part of his reason, I believe, is wondering who will maintain the graves when we are gone. The local cemetery association is a small group, mostly dependent on one woman whose father bequeathed the job to her. She hires someone to mow the graves every year, but there's never enough money for more maintenance.

May 30. A half-dozen Wilson's Phalaropes honk at the water's edge, sounding like ducks. I'd never seen these brown and white birds until a year ago when I noticed several of them spinning in circles near the edge of the dam. I grabbed field glasses and my identification books and learned they whirl in shallow water to stir up the aquatic insects on which they feed. Now I look for them every spring.

With Jerry's parents, we went to the cemetery on this Memorial Day and weeded my parents' graves and George's and decorated them with lilac blooms. I was crawling around on my father's grave pulling weeds while Jerry used the weed-whacker around the edges.

"Wouldn't my dad be proud of us!" I said to Jerry.

"Not necessarily. He'd say, 'You missed a weed there.'"

The older Hasselstroms I never knew have a cedar tree and a lilac bush in full bloom on their graves, plants now outlawed in the cemetery because their roots heave up headstones. I'm glad to see them there, creating shade and scent on that hillside where both are rare. Perhaps taking flowers to the cemetery is part of ancestral memory. Archaeologists say even the people we consider most primitive revered the graves of ancestors. I've always disliked funerals because they seem to demand expenditure only for show. Yet because my mother expected a lavish funeral, I used horrifying amounts of her money to put her in a pink coffin wrapped in pink roses: everything I deplored. But I fulfilled her wishes. Was my gesture intended to make up for my disagreements during her lifetime? This is not a question I can settle, even in my own mind.

Recently I read of a woman who learned at the death of a loved one that she could get a bargain on tombstones if she bought several at once. So she and her sisters bought tombstones, and she had a humorous saying engraved on hers. No one else would have approved, she said, especially after her death, so she decided to have that one thing about her funeral the way she wanted it.

"Life is a tragedy when seen in close-up," said Charlie Chaplin, "but a comedy in long-shot." I hope that I can keep the humor of the longer view as I study these journals.

May 31. I'm grateful to Jerry for overcoming my prejudices about four-wheelers. I use mine every day in spring and summer, zipping down to the garden to check on how the water is flowing or hauling the laundry up from the retreat house.

I woke this morning from either a dream or a revelation in which I saw my mother as she looked in some pictures when she lived in Pittsburgh, wearing a stylish suit, a wide hat, smoking a cigarette, and drinking a martini. She must have believed herself sophisticated, cynical and worldly wise. Perhaps she had grown disenchanted with the boy from home, the staid engineer she'd married. Perhaps around this time she was dazzled at a cocktail party by the smooth-talking salesman who became my father.

Mother had first done exactly as her society taught her to do: married her high school sweetheart after six years of courtship. Wisely, they prepared; he finished college and she finished secretarial school. After a wedding in their hometown, surrounded by friends and family they moved to the big city of Pittsburgh.

She said she "fell in love" with my biological father, Paul, so completely that she was willing to destroy her marriage to have him. In letters, he called her "Sweetie" and she called him "Honey Bunny." She married him two months after her divorce was final. Though they were married in 1938, I wasn't born until 1943. One photograph of her taken shortly after their wedding, though, shows her sad-eyed and forlorn. Was she pregnant when they married, and she lost a baby?

Later, when the charmer turned out to be an alcoholic liar, she didn't want to admit her foolishness to her parents or her friends at home.

By the time Paul came home drunk, towing me as a toddler, she was probably tired of his drinking and not working. That bloody dress, announcing betrayal by both her husband and a friend, ended her love. They divorced when I was four years old.

Soured on romantic fantasy, she may have been stuck with responsibility for a child she didn't want. I can picture her, belatedly, destroying the documents about her first marriage to conceal it. One divorce, from an alcoholic waster, was acceptable even in the late 1940s; two might seem promiscuous. This scenario fits precisely with her character as I came to know it. Excitement could make her so scatterbrained she'd say the first thing that came to her mind. But she could also lie deliberately and with forethought. "I lied like a good fellow," she said to me several times, smirking, telling me how she had hoodwinked someone.

I'm shaken by the storm of emotions my mother must have known and realize I understand those feelings because I lived through a similar story. I married after most of my contemporaries, waiting until I found the man I believed was the love of my life. The first time he came home smelling of another woman, I nearly shook myself apart with rage and pain. Chaos and destruction. But we were modern young married folks; we talked, we agreed on my forgiveness and his resolve never to do it again. He did it again and again and again until my love dissolved.

My mother endured and survived, and then was forced to relive the misery through me. Like her, I didn't want anyone to know how gullible I'd been, how foolishly I'd behaved when I saw that perfect marriage disappearing into the rear view mirror.

June

June 1. Here come the grasshoppers: three inches long, wearing military khaki, with swept-back wings like epaulets, legs sharp as boot knives, and jaws like tanker treads. Experts say a small hopper can consume twelve pounds of forage per month; larger ones eat nineteen pounds in the same time.

Riding to the garden on my ATV to move a hose, I outline a humorous article about survival on a ranch. First rule: keep your mouth shut when riding a four-wheeler unless you like the taste of flying insect life. Perhaps this is why ranchers are tight-lipped. Nothing like a sweat bee on the tongue to discourage chitchat.

June 2. Going to the garden is like visiting old friends in the cemetery as the grasshoppers multiply and leave only the remains of plants. The trellis Jerry built for the beans is bare; the hoppers have eaten the onion and potato tops. The peas are gone and the rhubarb leaves are a spider web of veins. Last year at this time I was harvesting peas and freezing rhubarb weekly.

The tomatoes in the raised beds at the house are doing well. I've moved herbs into the greenhouse and cold frames where I squash every hopper I can catch.

June 3. My mother's journal entries in the 1950s were filled with enthusiasm for ranch and community life, but in summer, my father often wrote in his journal, *"I stalked* [stacked] *some in eve."* He often went back to the hayfield after lunch and stacked hay until dark, much to my mother's annoyance because it meant they ate supper so late.

While my father was fulfilled by his work and his love of the ranch, doing mostly housework frustrated Mother. During the early years, she often mentions that they went to Grange meetings or a dance in Hermosa, but gradually that changed. My father disliked getting dressed up and would often come home too late for them to go to whatever outing she had planned. Did he do it deliberately? When Mother wanted him to go to town with her, he'd protest, saying, "But I've got work to do, Wife." She'd slam pots and pans around in the kitchen, scream and sob. At least once she dragged me to the car and slammed me into the back seat while she sat in the passenger seat and honked the horn.

He tried to keep Mother happy, but working until after dark was his way of keeping up with having too much work for one man. If he hired help, that caused trouble. One man stayed in our spare room for a week, but Mother didn't like cooking for him or having him in the house in the evening. So my father created a primitive bunkhouse. He moved the kitchen of an old ranch house to our place, had it rewired for light, and installed a wood-burning stove and a single bed and chair. The hired man had his own space to sleep and read, but he still had to come to the house to use the bathroom.

Then we discovered that though he worked hard all week, he drank weekends. One night he drove into the front pasture through the fence, and then out through a different spot. In the morning, his car was parked by the bunkhouse, festooned with forty feet of barbed wire. I think my father made him fix the fence before he left.

June 4. Blooming today: lilacs, yellow daisies, honeysuckle, flax, and currant bushes. I planted cabbage and Brussels sprouts and created a row cover of old sheer curtains to try to foil the hoppers. I've read that this year there are an average of eight grasshoppers in one square yard; they eat as much grass as a cow and calf.

Every day, I shuffle through pictures of my mother with me. In one, she is seated on a concrete step holding me as a newborn. Her hair is swirled elaborately on her head, and she is looking coyly over one shoulder, smiling as if she's trying not to smear her lipstick. Her dress drapes gracefully around her ankles and I am barely balanced on her knee, held in place with the tips of the fingers of one hand. Behind her are a screen door and the shadowy figure of a woman looking apprehensive. When I asked Mother who that was, she said, "Oh, we had a nanny at first, but Paul said we couldn't afford her."

A friend who looked at the picture pointed out that my mother was holding me as far from her as possible. She once told me that Paul wanted a baby, but she didn't. I avoid holding babies and have never cooed over them; I'm fortunate never to have been a mother.

June 5. My writing has gradually become my primary interest, sometimes to the exclusion of entertainment, friends, and other activities. What might my mother have wanted to do if she hadn't become a mother? She was 34 when I was born and as far as I know worked as an executive secretary both before and after my birth. Like most children, I didn't begin to see her as an individual until I was in my twenties. Perhaps she was tired of being a career woman by the time I was born but that didn't necessarily mean she wanted a child.

Today I was cutting brush behind the retreat house, piling up trash to get it out of an area we need to mow. I've worked hard cleaning up this area in the past couple of years, but today I found several pieces of some stationery that I gave my mother years ago. Puzzled, I recalled that the children of the former tenants played in this area. They had dragged many of my father's tools out here; I retrieved an antique Swedish saw ruined with rust and his favorite lariat. So probably they found some of my mother's stationery in the house and brought it out to play with. Yesterday, a dead tree half hollow inside blew down. Was the stationary hidden inside? And if so, how did it survive in such pristine condition? Why did it come to light now, as if it conveys a message?

Raking and piling brush, I keep looking at the house my father built for us when he married my mother, as if she's inside watching me through the picture windows she loved. She did what every good ranch wife was supposed to do: fixed the meals, kept the house clean, and followed my father's orders. When I was handling my father's probate I discovered that she was not a joint owner of the ranch; when we first met with the lawyer, she asked him how large the ranch was. If she objected to these arrangements, I will never know. Her journals are mostly a list of her labors.

My gift of stationery to her was a way of acknowledging that she had other interests, of encouraging her to keep up her connections with women friends outside this community. Since she had her friends in the bank investigate my father's finances before she married him, I find it strange she didn't demand more social life, but perhaps she tried and failed.

June 6. Just at dawn, I pick up the binoculars and look toward the calving pasture, brilliant green in the first rays of sun. A tawny shape trots across the field—deer? Pronghorn? No, coyote! He's big, moving fast, and his tan pelage is almost the color of the tall, cured grasses on the slope of the hill. He runs, head up, glancing once in a while toward the cows and calves. The cows begin to move away from him, but one cow turns and runs toward him, head up, bawling.

The coyote zigzags along the bottom of a shallow gully, ears pricked, then stops. He sits down, head slightly on one side, and remains motionless for several minutes. The cow has stopped bawling and running and stands, watching.

The coyote pounces on something, tosses it up, and swallows.

The cow bawls again and runs closer to the coyote, which looks over a shoulder, then continues eating.

The cow turns and runs into the shadow of a haystack, and I see a small black shape. She nuzzles it: her calf. She stands over the baby, facing the coyote, bawls again; the calf does not move. Perhaps sick or dead. I can't tell at this distance.

The coyote begins trotting up the gully, veering from side to side as though evading a sniper. If I look toward the cow for a second, the coyote vanishes into the bunches of redtop on the hillside. When I can no longer find his shape, I put down the binoculars and call Rick about the calf. Later, I see it up and nursing, so it's fine.

June 7. This morning I was reading my grandmother's 1911 recipe book from the Ladies' Aid. She handwrote recipes on every blank page, including her ice cream recipe: I made some for lunch.

Dandy Ice Cream

1 quart milk
Let come to a boil
Mix 1 pint sugar and ½ C flour together
Add to boiling milk
Cook two minutes, stirring constantly.
As it's taken from the stove, add beaten yolks of two eggs. When cold, add 1 quart cream, the 2 stiffly beaten egg whites, and 1 tablespoon vanilla. You need not have an ice cream freezer; just place ice cream in

the freezing compartment of your refrigerator for six hours or so. Because the ice cream sets up very hard, consider freezing it in small containers so it will warm more quickly to be extracted from the container.

While I stirred, I wondered if people, especially women, live longer when they express themselves creatively, whether the expression is in a journal, poems, quilts, singing in a choir in Rapid City, or even cooking. Mother wrote in her journals but didn't really do handiwork. She had a fine soprano voice and had been a star in the choir in our church, so she joined the church choir in Hermosa, expecting to sing all the soprano roles. But the choir already had a soprano; she was willing to share the spotlight, but my mother wasn't. At first, Mother went to Rapid often; she took me to swimming lessons. When I joined a 4-H club, Mother bought several sets of fancy luncheon plates for when we hosted meetings. Gradually these activities fell away.

My grandmother worried about my mother's depression and mental instability, which apparently had gone on long before that awful year of 1973. I've found letters she wrote through many years encouraging Mother to focus on others, not to feel sorry for herself. My mother wrote in her journal but didn't use it to analyze her behavior and change; I constantly evaluate myself, and worry that I will become as wildly destructive as she was. My father, by contrast, wrote of only the daily work in his journal and could not stand to let us—his wife and daughter, his closest family—see him cry for his dead sister. Was his mind healthier for not writing about himself? He deplored self-analysis, and perhaps he was right to do so.

Women writers, like ordinary women, are rooted in their ordinary lives; no matter how wonderful our accomplishments, most of us also clean the toilet. So we often write about practical, everyday matters, one reason why women's writing was dismissed as trivial for so long. Probably more women pass recipes on to friends than run for office, yet both activities connect us with the larger world.

Grandmother's journals were in her letters, her crocheting, her rag rugs, her recipes. Though she didn't write about her life, I find clues to her household's poverty and interests by reading what she has written or pasted into the blank pages. She records recipes for chow chow and mincemeat from green tomatoes, reminding me that with our short growing season and the poor soil at the Hey ranch, her family probably ate more tomatoes green than ripe. Judging from the recipes, she preserved everything: cucumber pickles, oysters, and suet pudding. Instead of buying readymade, she created her own mustard and catsup, taffy, Cracker Jack, peanut brittle and cream puffs. She noted that *"blueing is good for cleaning windows,"* and *"a teaspoon turpentine added to a pail warm water is good for cleaning purposes."* She even made beer, though I never saw her take a drink.

Tucked in the back of this recipe book, I found a note I wrote to her on Mother's Day 1957. She knew she could always find it there.

June 8. Blooming salsify shelters bees dressed to match: the blossoms are yellow and black.

"Scrawk! Scrawk!" We were sneaking behind the windbreak trees to see if the heron was on the dam; usually if we see it, we walk a different way so as not to scare it away. But today the big bird took off before we spotted it, its neck tucked into a tight S, flapping its enormous wings, squawking ferociously. A pair of red-winged blackbirds followed, alternately diving at its broad back and then veering away. Several times the little birds smacked into the heron—it lurched from the impact—and then veered away. Since herons eat birds as well as fish, snakes, and insects, perhaps the blackbirds are protecting their nestlings.

I'm still reading ancestral journals and letters every morning; today I wake up with the day's revelation.

If I'd had children, would I have become as angry and frustrated as my mother? Would I have had children if she hadn't been so angry all the time? Her anger never seemed to be at me, but at her unsatisfactory marriages. She not only had a child to care for without financial support from its father but she had to work and learn all the things she didn't learn growing up, like housekeeping and budgeting. When she found John, she made sure that he could provide for her, but he took her away from the city and the entertainment she felt she was entitled to after her long struggle to be a good mother.

No wonder she was angry. Instead of forgiving her, as I smugly expected to do, I am beginning to understand why she behaved as she did. I'm sorry I didn't learn these things when she was alive, but I can't believe it's too late for me to ask for her forgiveness.

At three I welcomed the new retreat writer and at six took a sandwich and a gin and tonic down to have supper with her. We talked for two hours about her writing, her hopes for the retreat, and all the other topics that seem to come up when a writer has decided my viewpoints can help her work. She lost her nervousness and relaxed enough so that I know the retreat will be worthwhile for her.

June 9. The light that flows across the hills as the sun sets is a rich blend of green and gold; the killdeer call *"kill-dee kill-dee kill-dee."* We can't sit outside because there's no wind and the mosquitoes come droning in from the dam. West Nile Virus has been detected in the state, so we don't want to be bitten.

I spent two hours in the morning talking with the retreat writer as we worked through my comments on her prose. Part of her writing concerned her grandmother, so our conversation inspired new memories for both of us as we compared our grandmothers' lives.

After lunch, I began writing comments on new work she gave me to critique. Absorbed, I was slow to realize that the dogs have been barking wildly for a long time, not their usual, "We saw a rabbit hop past" bark.

Outside the dog pen, I see varied birds—Eastern kingbirds, robins, red-winged blackbirds—flying around a juniper tree just south of the house, screeching.

I approach cautiously, knowing there is a robin's nest in the tree, and push a branch down for a better view. Six inches from my face, a bull snake is coiled around the tree trunk, half its body lying across the nest. Beneath its coils, I can see a wing, the body of a parent bird. The snake's body bulges in several spots with the babies or eggs it has swallowed. The snake raises its head and looks at me, tongue flicking in and out of its mouth lightly.

Nature is at work. I call the dogs inside.

June 10. When we walk the dogs in the morning, the sun lights a single drop of dew at the top of each grass blade. We spot a coyote on the hillside. Below, the cows move through the fog spread like whipped cream over low hills and gullies. The cows' heads are up, their bodies black rectangles, and their legs disappear into the white foam.

Then a calf bawls and suddenly forty cows are running toward it, bellering, ready to save it from the coyote they can smell. The calf runs among the cows until he finds his mother, then begins to nurse. A few cows buck and leap, twisting in air, pirouetting on their front feet. They are so healthy they have extra energy to spend playing.

At the first bovine bellow, the coyote stops, ears up, looking toward the herd, then vanishes.

We're eating radishes and lettuce from the garden every day, sharing with the retreat writer. Tonight Jerry grilled ribs; I fried potatoes and made a huge salad. A light breeze mixed with the smoke from the grill let us spend a few minutes on the deck enjoying the stars without mosquitoes.

June 11. I took the final batch of handouts and written comments to the writer at eight this morning and hugged her as she left for home, smiling. While I stripped the bed and emptied the refrigerator, I thought of her driving home as she makes plans about her writing. What a wonderful job I have created for myself!

Later in the morning, we are standing in the auto license line on the second floor of the Rapid City Courthouse when I hear the *chock chock chock* of a woman's high heels on the marble floor.

Suddenly I am six years old, following the sound of my mother's high heels across that same marble floor, running my hands along the smooth railing as we go up the marble steps. Why were we there? My mother had no car while we lived in town, but I clearly remember looking down from this second floor balcony to the brass dome over the front door, wanting to climb over the marble railing to touch that shining brass.

Those footsteps remind me of Mother in her broad-shouldered fur coat slipping her galoshes over her high heels, setting her black felt hat precisely on her head and waving goodbye to me as she clomped down the snowy street toward the bank where she worked. My grandmother must have been living with us at the time. Somewhere I have a photograph of Mother in that outfit, but I'd forgotten until I heard those high heels on marble.

What memories hide in our brains, never released unless we hear the right sound, or smell the precise awakening scent?

June 12. I'm making stock: beef scraps simmered on the stove all day with onions, garlic, and vegetable peelings.

My journal for this date in 1987 tells me I had a migraine so bad George took me to the emergency room at the Air Force Base, where his retired status enabled us to obtain care. The attendants put me on a gurney and shoved me into a room with lights so bright I screamed and passed out. When the doctor came, he gave me a Demerol shot and sent me home to bed. The next day we moved cows and calves over east, and my father said I must not have been too sick after all.

After George died, one of my neighbors came to the door one day when I was in bed with a migraine and pounded on it until I lurched out of bed and opened it. I explained that I had a migraine.

"I never have headaches," he said.

I managed not to say, "Because there's nothing inside your skull to ache." He typified the local attitude: a headache certainly isn't serious enough to make a person sick.

Looking through the journal for the rest of that year I realize this was the year when we went on the buckskinning ride from Cody, Wyoming, to the rendezvous. During the trip, my rented horse died, and it was still the most relaxed either of us had been for months. A month later, we had the great confrontation with my father, where he insisted he'd paid for our house, that he never intended for us to be partners, that we got more out of it than he did, and on and on.

Finally he said I was a worthless bitch.

George leapt up from the table and roared, "You don't talk to my wife that way. You get out of here before I throw your skinny little ass down the steps."

We spent the rest of the summer making calls and applying for jobs, yet we still did most of the ranch work. My mother cried every day. My father seldom spoke to us. I knew his mind was not right, but I could do nothing.

In December, my parents went to Texas as usual, leaving us to care for the ranch. My father's letters began *"Dear Child,"* and were signed *"thinking of you,"* as if the whole rupture had not happened. Still, George and I were determined to find a way to leave the ranch. We were getting along better as we planned our getaway.

At dusk, a pair of eared grebes dive and swim on the pond. Nighthawks twist the strands of darkness, swinging in great loops over the prairie. Their cries and booms seem to echo, bouncing off the grass and sky, the plaintive *peent! peent! peent!* before the throaty boom of their killing strike at some mosquito or fly. What we call a nighthawk is really a nocturnal nightjar, one of the best insect-eating birds we have.

June 13. I've been hand-killing grasshoppers in the garden; they're so big some of them are easy to catch. Pans of water containing molasses drown them by the dozens. Still, millions of hoppers a quarter inch long are eating hollyhock leaves to lace, so I'm spraying with garlic and mineral oil.

Another retreat writer comes today. I'm glad to stop looking at family journals to concentrate on someone else's writing.

June 14. Each evening, if the breeze is strong enough so we can sit on the deck, the nighthawks entertain us with their cries and acrobatics. They are my favorite birds, and not only because they are eating mosquitoes while they make loops over our heads.

One June day when we lived in Cheyenne, we were walking our dogs around a man-made lake in the city park, watching ducks floating serenely. With a whistling sound, at least fifty nighthawks roared over the trees at the lake's far end and swooped along the water, flying over and under and around each other. At the near end of the lake, they spiraled up, flipped over, and dived in a half-dozen arrow-shaped bunches directly toward the ducks. Nighthawks eat bugs, not ducks, but some flew so low we saw them strike ducks on the heads. The ducks squawked, flapped, splashed, and scattered. Again and again the nighthawks dived, shrieking. We struggled not to laugh at what was clearly harassment.

June 15. The rock garden is gorgeous with blooming flax. I placed a dozen aqua glass insulators from the old telephone poles in a long curve down the hill, so they wind like a stream through the sandstone, agate, quartz, and other stones. They sparkle like water in the rain, and also slow the flow of runoff down the hill.

This week's retreat writer has been writing and revising furiously, so I spent much of the day reading and commenting on her work. I'm inspired by her energy, but only have time to take a few notes on my own work.

June 16. After the writer departed, I helped my assistant, Tamara Rogers, change the sheets, tidy the house, and empty the refrigerator, since another retreat begins this afternoon. We left doors and windows open so the house blooms with the scents of new-mown grass and sun-struck pine trees. For the traditional "welcome" bouquet, I found a wild yellow rose.

At sunset, I walked the three visiting writers around the retreat house in our usual orientation stroll, then invited them to join us on the deck after supper. We had just enough breeze to keep the mosquitoes away, and a great view of low-hanging, pendulous clouds that might portend a storm. While we visited, the prairie went about its business: ducks paddled on the pond, then upended to graze along the bottom; killdeer paced the shallows; frogs leapt; meadowlarks, blackbirds and barn swallows zinged past, hauling bugs to their nests; the cattle lay in scattered groups, digesting: all oblivious to our human presence.

June 17. The sky at 4:25 A.M. is the color of a black eye. Five antelope—one buck and four does—are silhouetted on the top of the hill. While I watch and write, a whitetail doe stands up from the grass nearby. The pronghorn raise their heads and mill around; the buck thrashes his horns against a tall weed.

The buck advances on the whitetail; she backs away. All five pronghorn surround her and she trots off, but looks back at them. The buck wanders off but the other pronghorn keep sniffing the ground, gathering in a circle, then moving toward the deer; they remind me of teenage bullies harassing the nerdy kid on the playground.

Is the doe trying to lead the antelope away from a hidden fawn? When I go outside and call the dogs from their pen, the antelope take off and the doe disappears down the draw.

Three writers keep me busy writing comments, but fortunately they had sent their work well in advance, so I got a good start. They are comfortable enough with each other that they pass their writing around, and our group discussion is exhausting, but exhilarating.

June 18. Thunderstorm with a quarter inch of rain; the cows drift east ahead of the wind. Our rain tank runs over. All day the cattle graze and move gently around the pasture. Sometimes a cow lies or grazes with a dozen calves around her, babysitting while her sisters go to water. At dusk cows and calves bawl, finding each other by smell and voice. The calves nurse and then play, butting heads, running wildly around the hillside. Gradually they quiet and curl into sleep.

This was another busy day of writing at the retreat; I went back and forth several times with handouts and comments.

June 19. Shafts of deep blue and purple penstemon stand erect in the rock garden with bursts of orange and yellow and brown gaillardia blooms. So many penstemon varieties exist I can't keep them straight but these are mostly the common Rocky Mountain. Before sunrise, yellow evening primrose blooms open like silk parachutes, then snap shut, shrivel and drop off by sunset.

After the writers left, I collected the laundry. While it washed and dried, I went through my journal, expanding the notes I took in the hurry of the past few days. Teaching this way is tiring, but much more satisfying than standing in front of a classroom.

June 20. An antelope doe stands up on the hillside, looks around. Then one—two—and three fawns leap up. Two run around the hill, making big loops up and down, chasing each other. Their legs are so long, their bodies so small, that they look like daddy longlegs. Two other does in the field below the hill raise their heads and amble uphill; apparently antelope practice group fawn care just as cows do. As dusk falls, they all fade to invisibility. The sun is a red gold ember buried in ashy clouds.

June 21. The Summer Solstice is the year's longest day, and yet the day when summer begins its decline into winter. Growth leads naturally to death and gain to loss but these are natural matters, not for despair. A mariposa lily blooms at the curve on the trail to our house, triggering memory. "Sego lily," my father always said when he brought only one—because they are so rare—to my mother. The sego lily reminds me of their quiet joy in each other, how he preserved the plants, and her gratitude for his thoughtfulness.

I have watched all day for the antelope fawns; I've never before seen three together. It's good to be reminded that everything I've seen in sixty years on this prairie is only a fragment of all there is to see and know here. No one will likely ever know this piece of the plains as intimately as I do, and yet I know so little. A lot of what I know will vanish when I do, lost as my brain and bones dissolve into dust. Those who follow will know it in their own way, but I worry about what that knowledge might be. If my rancher neighbor's son eventually owns this place, he'll know it in a lot of the ways I can approve. If a subdivision covers this land, the loss will be unfathomable, not only for me but for others.

June 22. My father's journal for this date in 1979 says, *"Linda had front page in RC Journal about woman stealing her writing."* To me he said, "Why would anybody want to steal that junk?"

The story was written about my lawsuit against the romance writer who, as the court ordered me to phrase it, "committed copyright infringement" by using passages from my published journal *Windbreak: A Woman Rancher on the Northern Plains* in a couple of Harlequin novels. The newspaper printed paragraphs from both my work and the romance writer's, making the similarities clear. She's still publishing romance novels. I wonder how closely her editors check them.

I've always lacked confidence in myself, but writing had become my salvation, so my father's pronouncement shook me; I'd always assumed he was proud of my writing.

June 23. I've turned to my father's letters to blot out bad memories with good. In his letters when I was in college, he expressed more love than I remembered.

"At the Breakfast table, Monday Nov. 2, 1964," he wrote. *"Thot I'd better give you my Monday morning advice while it's still fresh in mind. Hated to say goodbye to you the other nite on the phone but know it was getting expensive. I guess we had better call oftener and make them shorter. About the car, hope it gives you good service for a while—perhaps you shouldn't ever drive it over 55 miles an hour. On any car and especially an old one the costs and the danger seem to increase at much faster pace than the speed of the car."*

The car was the 1954 Chevrolet he'd helped me buy for high school. Since I was working at night, I must have mentioned being tired because he continued, *"Keep good tires under you—just forget all about this business of being tired or sleepy or blacking out or anything else when you are driving. Remember that sounds good in books but if a person is dead what difference does it make what people read. If you get sleepy—tired—dizzy or anything else—just stop driving. Don't go to school—don't go to work—go to bed, then get up and do things right. Remember cars just aren't interested in alibis or excuses. Don't be angry with me sweetheart—but I know how easy it is to slip into things like this especially when a person lives alone."*

Such wise advice, and he'd lived alone much of his life; he was 43 when he and Mother married. Now, twenty years too late I wish I'd asked his advice about living alone when I was divorced and again when I was widowed.

June 24. By 1987 I was teaching at a college in Rapid City, and George was taking adult education courses at night for his MA degree as we made plans to leave the ranch.

My father wrote in his journal that he got lost in town going to an equipment store where he'd been hundreds of times. He described people in his journal to help him recognize them. Both he and Mother complained that I never told them where I was going so I left a note on their dining room table as I left for work every day. The notes disappeared into the mess of newspapers, grocery lists, and letters, so when I stopped on my way home from work, they would be angry that I'd "gone off without telling us where you were."

June 25. The Say's phoebe slips under the deck to feed her babies so deftly I hardly see her even though her nest is just above my study window.

A thunderstorm struck this evening, violently whipping rain at us from the southwest, pounding my tomatoes in the raised beds. Lightning was frequent and close, slamming the hillsides and splitting the air in all directions. A few small shards of hail fell, but nothing like the baseball-sized hail that pounded Rapid City last night. After the storm, the rain gauges held only six-tenths of an inch.

Just before we went to sleep, we thought we smelled smoke. As usual after a storm, I turned out the lights and went to the deck to look around for any sign of flame. I looked at the retreat house and barn and toward the stack lot where piles of round bales wait for winter use and toward the field where Rick had been haying that afternoon.

But we weren't really worried. Too wet to burn, we said, because the pastures are full of green grass.

We'd both spent busy days. Jerry mowed around our buildings in the morning while I worked with other Hermosa Arts and History Association volunteers sweeping and rearranging furniture and display cases on both floors of the massive building in Hermosa. I love this story about our community: for ten years, members of the association have been renovating the 1895 school building to become a museum and community center.

In the afternoon, we went to the Great Plains Native Plant Society open house at the botanic garden in my front pasture, where a botanist found a native plant new to most of us: Fame flower, a succulent so tiny we could only see it on hands and knees with a magnifying glass.

Still, we were both awake at three with aching sinuses from the change of weather when the dogs barked. Looking out the living room window, I saw headlights and leaned closer to the screen to see when a voice spoke in my ear. "Linda; this is Bill Brazell. There's a fire." I jumped a foot and called Jerry, then opened the door. The time was 3:15.

Let me be clear about what my neighbor had done. He woke about 2:30 and smelled smoke. He got up, looked around his house, and determined that the smoke was coming from the east, toward my house. He couldn't remember my cell phone number, so he roused his wife, got dressed, got in his pickup and drove down his ranch road toward the highway. When he topped a hill, he could see the fire and knew it was on my property, a mile east of my house. He guessed it was a shed or hay bales, so that with the wet ground and grass, it was no threat to anyone's home, but property was burning. He drove the muddy road to my house and was about to knock on the door when I looked out the window. He spoke quietly so as not to scare me. He knew I couldn't see his face in the dark. Neighborliness in the country is not simply living next door to someone.

I could see flames flickering east of us, but couldn't tell if they were in a stack lot or in the cattle shed that stands on the other side of the tracks.

I called Rick and we decided against calling the fire department, figuring the fire was in a haystack and no amount of water could put it out. We might keep it from spreading to other stacks. Jerry and I grabbed shovels and tools and headed east, while Rick got on his tractor and headed toward the fire from his house north of us.

A fire unit from the Hermosa volunteer fire department caught up with us, called by someone who saw the flames from the highway. We all wallowed through the muddy field and pasture roads until we could see that the fire was burning bales east of the railroad tracks. The firefighters left.

Rick used the grapple fork on the tractor—like a large hand—to move bales that hadn't started burning. Each bale weighs about nine hundred pounds and on some trips he moved two at a time, stacking them away from the fire. Eventually he'd isolated about ten burning bales in the center of the hay yard.

By that time, a fire unit from Fairburn had arrived. Rick was driving the tractor to the burning bales, grabbing one at a time and backing out onto open ground, where he'd scatter the flaming hay as quickly as possible. Then he'd go back for another.

With the neighbors on the fire truck, we started using shovels and pitchforks to break up the burning bales so they'd burn faster. Each time Rick dragged a bale away from the burning pile, he'd turn and drive through the bales he'd brought out earlier, breaking them up as the smoke swirled around the tractor. He had to keep moving fast so the hoses on the tractor's hydraulic system wouldn't burn. He couldn't see much through the rolling yellow smoke, so we stayed out of his way. Each time he drove through a burning pile of hay, the speed of his passage sucked the smoke after him, so it looked as if the back of the tractor was on fire.

I became aware that the smoke smelled just like drying hay: sweet, not smoky at all. During breaks, we introduced ourselves to the people we didn't know, leaned on our tools exchanging neighborhood news, and reminisced about previous fires. Our area fire departments are now dispatched by a call to 911, but most are still volunteer forces, so neighbors are always helping neighbors. In addition, anyone who finds out about a fire in the area jumps in a truck and comes to help. While we were working on this fire, one of the men got a call from a neighbor who had been called by another neighbor who was on his way to work in Rapid City and saw the flames.

I've learned that newcomers don't realize that the fire departments depend on volunteers. They dutifully pay the fire assessment on their property taxes, and sleep tight while their neighbors put out the fires. The volunteers are aging: many of those from the surrounding towns are more than seventy years old. Rural fire departments are having a hard time recruiting younger people, who don't want to invest the time required for training in fire-fighting and emergency medical work. The volunteer fire department in Spearfish, a town of 11,000 in the northern Black Hills, has disbanded. If a town that size can't keep a volunteer fire department going, how can these smaller communities cope?

Finally the Fairburn truck left, but we all kept tearing bales apart with our pitchforks and stamping out flames; we headed for home at eight that morning and by ten the fire seemed to be out. The wind was blowing, so we'll watch all day for flare-ups.

No doubt lightning struck the bales early in the evening, and the fire smoldered through the storm, including the half-inch of rain, until it got enough strength to burst into open flame. We still don't know who called the fire department.

Today I'll write my usual summer checks to the volunteer fire departments in the area. Once again neighbors, both on the fire trucks and off, saved a rancher's property, knowing next time he might be saving theirs.

June 26. Today I counted blooms on the hillside: mariposa lily, sweet William, beardtongue, pincushion cactus, bracted spiderwort (which my mother called "cowslips" so I can seldom remember the real name) and salsify or "oyster plant," a description supposedly given by pioneers from the flavor, which does not recommend it to me.

The business of living consciously takes a lot of time. Today we hauled recycling to town, piling the bags of glass and cans in the back of the pickup with sacks of newspapers and piles of cardboard. I loved pulling into the recycling site to see half a dozen other older people hauling stuff out of their cars, all smiling ruefully at one another. In fact, most of the people I see at the recycling bins are over sixty. I hope this is merely coincidence.

We recycle energetically with compost bins here and at the retreat house, to teach writers the benefits of compost. I put scraps of meat and bones with leftover vegetables into bags labeled "Beef" and "Chicken and Pork" in the refrigerator freezer. When the bags are full, I dump it into the stockpot, adding water, herbs and spices to make stock for soups and stews.

June 27. A hundred degrees at noon; flies crowd every patch of shade; cows stand belly deep in the pond. The calves beside them show only their heads, scattered across the surface of the water like black and white lilies.

Thinking about my family's journals, and mine, and how they correspond with my memories leads me into deep water sometimes. I sometimes experience memories complete with sensory impressions, but some dreams make no more sense than the fright house at the county fair. I dream through situations that seem to have no bearing on my daily life or past—and yet my mind generates these scenes. I know some people create believable falsehoods when the life they are living does not suit them. My mother, for example, decided that my father's death was a blessing because "he'd been so sick." She could not believe that his mind had been clouded, even when he disinherited me; instead, she substituted the fiction that he was physically ill and death was a release. Her mind created an alternate story that helped her accept widowhood.

Am I mentally rewriting my story to make it more acceptable? Is my writer's imagination falsifying the facts? As I read my parents' journals, I remember that I am reading the truth they believed at the time. Perhaps they were revising the story to make it more palatable.

June 28. Larkspur, iris, and wild pink roses bloom on our hillside and down the entrance road. Last night several people drove into our road, relieved themselves and used large quantities of toilet paper to tidy up, and then scooped an amazing quantity of beer bottles and cans out of their car onto our property. Now I have more understanding of why "NO TRESPASSING" and "PRIVATE PROPERTY" signs have proliferated here. We didn't used to need those because almost everyone understood and respected privacy in property and person.

Lately I've read letters to the editor blasting ranchers for having "all that space" that people aren't allowed to play in, not understanding that we use our private property to raise their food. When I find bags of household or picnic trash dumped in our driveway, I search it with care (wearing gloves) hoping to find an address, knock on the door, and politely return the trash to the people who dumped it here.

Once, while traveling in a national park, I found a bag of trash discarded along a beautiful river. Inside I found letters from parents to their college-age children, on vacation in the park. I wrote a note saying where I had found the trash and mailed it, as cheaply as possible so it would have time to ripen, to the parents. I didn't include my address, so I have no idea how the story ended. But imagining the ending makes me smile as I pick up more used diapers.

June 29. On this warm, blustery day, reading my 1982 journal, I remember rain and fog as I rode my horse Oliver over east with George, my father, and a neighbor to move cattle from the 640-acre pasture into the south pasture. Then we cut eighty cows and calves into what we called the "government" pasture—BLM land we leased—and checked fence.

"In spite of the cold," I wrote, "the ride was great—the rain so cold and fresh-tasting, air so sharp and full of grass scents. Riding in rotten weather is really a test of one's ability to do it and enjoy it. The horse was cold and wet but he enjoyed it too, being more careful of his footing than usual because the rocks were slick with rain."

That passage brings it back: standing in my stirrups, licking rain from my lips and blinking it out of my eyes as we hazed the cattle along. When we finished, we turned the horses loose, threw the saddles under a tarp in the back of the pickup, and crowded into the front seat with the heat up high. The smells of our own and the horses' sweat and soggy, dusty denim filled the cab as we drove home, smiling and strong and young. We passed around an old porcelain jug wrapped in damp burlap, the water inside tasting sweet with a hint of sage. Never again.

June 30. Last night after a thunderstorm I looked east and saw red taillights, then a set of headlights. Watching through binoculars, I realized several pickups had gathered on the Missile Road north and east of us, perhaps watching for fire caused by the lightning strikes. I assumed these were neighbors with legitimate business there, so I went to bed, confident they would take care of whatever the problem was. This morning I learned the wind had blown down a power pole and started a fire. Volunteer firefighters and neighbors quickly doused it and then stayed there for an hour, visiting between the pickups, watching for more fires. What a fine way to socialize while performing a community service.

We've been back on the ranch nearly four years and today I found the journals I made to facilitate our move. I created photo-journals of the house where we lived in Cheyenne and the two houses on the ranch, both the one where I grew up, occupied by tenants after my parents died, and this house I built with George. We had decided that my parents' house would become the retreat house and we would move into the other one. To help prepare, I photographed and measured each room and all our furniture. As we traveled back and forth, we gradually decided what to keep and what to give away or sell.

I remember one awful day of exhaustion when I was cleaning out my mother's house after moving her to a nursing home. The tenants had moved in and a woman I did not know was following me from room to room asking questions as I tried to decide what to take to my mother, what to discard, and what to keep. I opened a cupboard over my head and tangled, crumpled papers poured over our heads, landing ankle deep around us. Some were letters, but I shoved them into black garbage bags and threw them out. Only later did I realize that many of those letters had been from my grandmother. Of course I'd give a great deal to have that moment back. Regret. What a wasteful emotion.

— July —

July 1. At dawn, a cool breeze ruffles in the south windows, lifting my hair. These are typical rangelands summer days, the grass and gardens growing, cows and antelope and deer drifting over the green hills, grazing and fattening. A raft of ducklings floats on the pond. If we were ranching, we'd be busy putting up hay in the middle of all this relaxed natural abundance.

At dusk, hummingbird moths as big as my hand, nearly invisible in the gray dusk, flutter among the evening primrose. Stalks of blooming yucca, sensuously curved, cast long shadows in the dusk.

I've finally finished arranging my grandmother's letters in order from 1928, when my mother went to secretarial school in Nebraska, through 1977; she died in 1980. Reading the letters brings my grandmother's voice to my ears again.

Consoling her daughter on the breakup of an unsuitable romance, Cora wrote, *"I always thought you would soon see how foolish it was to want to go with any old thing that came along but we all have to learn that by experience for no one can tell us and make us really understand till we know for ourselves. . . . Dad got quite a kick out of your letter he says in case he doesn't get you a room built on before you come home to settle down to be an old maid he may be able to fix you up with an apartment in the basement that would be a nice warm place for your canaries and pussy cats yes we really expect you to be an old maid especially that you are worrying about it at 20."*

My mother later admitted she was "boy crazy" and prone to dramatize every situation just as she was when I came into her life.

In those early years, Cora was cleaning houses and working in a café as well as in her husband's auto shop. She was afraid she would soon be *"clear out of extra work then we will miss it,"* yet she sent my mother five dollars every week and sewed dresses for her. I wonder if those dresses were stylish enough for my mother, even at a school in Grand Island, Nebraska. She always liked the latest fashions, and photos from that period show her in skimpy flapper dresses, typical 1920s attire that I doubt my grandmother made.

In another letter, Cora writes to my mother, *"I would just feel ruined if you got kicked out of school so be a nice laidy . . . that is all that it takes to get one in the best of company and make people who do a mount to something respect and love you. . . .God bless my dear little girl and keep her from all harm is prayers let your konchance* [conscience] *guied* [guide] *you and you will never get in bad I am sure the main trouble is we don't think things over seriously or enough before hand."* I wonder how much her parents knew about my mother's wild behavior. In 1929, Grandmother wrote, *"It does seem that you and your jentlemen friends just can't get a long."*

Grandmother's letters rarely showed her as depressed, but in December she wrote, *"I hope you don't plan to spending but very little for Christmas presents this year we just can't as we have so many things to pay by the first of the year it shurley would seam good to have everything paid and be out of debt."*

And, *"I hoop your boy friend proves to be as nice as he seams don't let him rush you to strong so you can't get sufficient sleep to do your work I am so tired and feel so bum I don't know just what to do."* When my grandmother cared for me in Rapid City while my mother worked and dated numerous men, she was always cheerful and supportive; now I know she was remembering my mother's history.

July 2. This morning we drove into the Hills for breakfast at a tourist lodge. On our way, we stopped for a hen turkey followed by nine babies who trotted across the highway, followed by a lone turkey who pranced across as if she hadn't a care in the world. After we passed, we looked back to see her babies trying to catch up with her, scrambling on the asphalt, dodging cars, but they all made it. Overhead, we saw a heron, followed by a golden eagle and a couple of turkey vultures. As we passed a field with tall brome grass, a doe raised her head, then looked down to her fawn, which was trying hard to see over the grass. Summer's plenty made flesh.

I was leaving the garden with lettuce and radishes when long fingers of rolling white reached across the sky under low gray thunderclouds. I dragged a few potted tomato plants under the deck. As the wind hit, I stepped out the upstairs door to save another plant and had to grab the railing to keep from being blown over. Stinging drops of ice blasted my face.

Inside, I looked at the clock: 2 P.M. For perhaps five minutes jagged chunks of ice fell, some as large as a chicken egg. The wind seemed to lighten for a moment, and I thought, as I always do, "Well, that wasn't too bad; maybe the garden will make it."

Then the wind roared and chunks of ice the size of mothballs pelted down, sounding like strings of firecrackers. The glass in the windows on the north made cracking sounds, so I collected towels and began mopping water driven through the window frames. I couldn't see beyond the driveway. When the racket stopped, it was 2:17 P.M. I called Jerry. He had taken refuge in the tin-roofed blacksmith shop, stuffing cleaning rags in his ears because the sound of the hail on the roof was about to deafen him.

Two and one half inches of water stood in the rain gauge, but it had shattered at that level, throwing pieces twenty feet away; neighbors to the east recorded three inches. Sweet clover stood four feet high an hour ago, its stems a half-inch thick; it's chopped level with the ground now.

The water in the dam rose a foot in minutes. Floodwater sloshed through the corral, moving manure to the fields. Yesterday a hired hay crew mowed several hundred acres of hay, the swather rumbling all day long. I mumbled that my father never laid down that much hay at once because of the risk of it getting rained on, and grumbled about younger ranchers who deplore our cautious ways. They spend a lot of money to buy big machinery so they can mow more hay faster. Well, today all that hay they knocked down is soaked, probably ruined. They may salvage some hay if they rake it and turn it over to dry several times, though it will lose nutrition. Drifts of hail stand six inches deep.

July 3. Surrounded by the smell of rotting vegetation, we survey the hail damage: tons of plants that should have been hay and forage have been shredded to mulch and fertilizer. The boards in the redwood deck have survived twenty years of weather, but they are pocked with raw new dents. The rhubarb and potatoes have vanished. The grasshoppers are still chewing vigorously on whatever remains but I hope some of them were killed by hail.

July 4. On July 4, 1982, my father wrote in his journal *"worlds of grass"* but today the grass that looked so good yesterday is flattened. The native grass will spring back, but the alfalfa will have to grow from the roots. Today the hay crew is out with rakes turning over the hay they'd cut before the storm. They will work through the holiday. We have no family gathering planned, but we may watch fireworks in Hermosa tonight from the cemetery hill.

The pond looks scummy, reminding me why my mother didn't want me to learn to swim in the pasture ponds. She insisted on driving me to a pool in Rapid City where I took lessons while she shopped. One day, driving too fast, she came up behind a truck carrying steel pipes.

"You're going to hit those pipes!" I said, or possibly screamed.

"Shut up!" she screamed. I dived to the car's floor and curled into a ball just as the car smashed into the truck's rear bumper. A dozen steel pipes—carrying an orange warning flag—punctured the passenger side windshield where I'd been sitting.

My mother shrieked until someone got her out of the car. Curled up on the floor of the car, I kept my eyes shut while I shook glass out of my hair. I finally raised my head enough so someone saw me and opened the door to help me out. Mother told my father that I made her too nervous to drive. I vowed never to ride with her again, and my father usually drove after that. I never went back to swimming lessons. I don't swim very well.

This morning Jerry was reading me an excerpt from a magazine article, and we wondered what the word "skirl" meant. When I looked it up in the big dictionary we keep in the dining room, I realized my parents kept a dictionary on the buffet beside their table. One of my jobs as we discussed the news over meals was to look up unfamiliar words. Besides learning how to spell and define words, I learned to enjoy a lively discussion that might involve disagreement; we'd resort to the dictionary rather than yelling.

July 5. Every morning as I sit in bed with my journal, I glance at my "Grandmother Corner," where I hung five of her crocheted doilies in frames. I can still see her, alone in her little house, crocheting while thinking of her children and grandchildren. In a collage, I framed photographs of her at different ages: at her wedding; smiling with my mother, who has persuaded her to wear knickers; feeding her chickens at the Hey ranch against a backdrop of snow; and sitting in her favorite chair reading. Arranged around the photographs are an Eastern Star handkerchief, because she was a fifty-year member, a large hair pin, her biggest crochet needle, a buttonhook for her shoes, and a curling iron she heated on the wood stove to create ringlets for both my mother and me.

In the evenings, we've been watching from the deck two batches of ducklings on the pond, a dozen each. Tonight, a single hen appeared with five ducklings. We've seen one antelope doe and her calf and wonder if the vicious hail killed the other ducklings and two of the antelope fawns.

July 6. The early hatch of red-winged blackbirds are learning to fly, gathering on our gravel entrance road to practice takeoffs and landings. They flutter along in front of the car, bouncing to an ungainly stop in the middle of the road. I brake, then edge to the left. A male red-wing flies across my windshield, shrieking. The teenager flaps its wings and lurches left. I stop. The young bird runs a few steps and takes off, flying ahead of me, straight down the road, very s-l-ow-l-y.

I ease the car forward. A second and then a third teenage bird appears from the deep grass and flaps into the air.

The first bird looks over its shoulder at the monster car, loses altitude and flops ungracefully to the gravel. Two sets of red-winged blackbird parents sweep overhead, screaming at the juveniles, flashing past my windshield, doing arabesques and pirouettes the youngsters must envy.

Bird by bird, foot by foot, I drive slowly until the young birds began to realize they can land somewhere else.

In July 1974, I wrote in my journal: *"All summer the black thunders growled from the ridges but came no closer, as if there were a hole in the sky above us."* Every day the temperature is above 100 degrees and at night rarely falls below 70. We sleep with the windows open but the wind is hot and smells of rotting vegetation from the hail; the highway noise blasts us. When our outside work is done, we sit in the basement to read, play with the dogs, not talking much.

July 7. Mist rising from the pond. Strawberries for breakfast. I need to recall and write about the little memories of Mother and Grandmother. I won't expect any revelations but want to see where fragmentary recollections will lead me.

My grandmother talked about how she and her sister-in-law Edith "went to housekeeping" as newlyweds in a one-room log cabin with their husbands.

"We was easily made happy in them days," she said with a smile. "We just dug in and worked."

Busy day: another retreat in progress, and the writer needed considerable casual conversation to become comfortable before I could begin to discuss her writing.

July 8. My father was born on this day in 1909. In a letter dated July 12, 1966, he wrote, *"Much too late at the breakfast table. I want to thank my dear girl for those birthday things. I've said before and I say it again—I wouldn't trade my girl for any other—it just shows what a person can do with a good product if he trains it correctly. Anyway it was wonderful of you to do it and I don't forget easily."* He couldn't restrain his wry sense of humor even while being sincerely grateful.

At 23, I was married to my first husband in March of that year and was living in Missouri. I probably sent my father homemade bread and perhaps some books.

I've finally reached a kind of calm, more than two decades after my father's death, as the memories of his irrational behavior dim a bit and I can remember his many good points. For so long he was such a good father, the only father I ever knew, so I have hated recalling the bad times.

How did he learn to be so kind and yet so authoritative? I assume he was emulating his father, or perhaps his mother. I never asked him or any of his brothers or sisters. Of course now that they are all dead I wish I knew.

July 9. Eleven tree swallows are lined up on the electric wires above the closest bluebird nest box, chattering and preening. Audubon guides say tree swallows often nest in bluebird boxes like ours. All day long we watch parent birds flash past, duck into the door, stuff insects down peeping mouths, and zip away for more. Sometimes one bird waits while the other feeds. My reference book says the birds forage in flight from dawn until dark, plucking insects out of the air, and that males may tend two different mates in separate sites, which may explain some of the bickering we have seen around these nests. We get close views, because the house is an obstacle in their flight paths; on the deck we sometimes have to duck as they skim past.

In a twittering flock, the tree swallows leave the barbed wire and spiral up toward the clouds. Just because they can.

Dragonflies hover in the yard, their wings clicking as they gobble mosquitoes. We saw a bull snake in the grass on our walk; it spun around to face us so we could admire its glossy black and brown markings. As the snake straightened out to leave, Jerry laid his tape measure alongside: five feet.

July 10. Rick spent the morning raking hay for the second time, trying to get it dry enough to bale, then stopped in before he went to lunch. While we drank coffee and visited, a thunderstorm rained and hailed, soaking his morning's work. He quoted my uncle's saying, "Rain makes more hay than it ruins." Surely he was right; the rain was soaking the hay, but also producing tons of grass. We agreed that the cattle would gobble up the damaged hay on some day when it's forty below zero. Ranchers have developed a rationalization for every disaster.

July 11. Officially today, the humidity is 51 percent, but we are used to the dry heat of 100 degrees, so, the air feels like a wet towel. We had the wettest May in recorded history. Low places that are usually cracked mud at this season have several feet of water and mud in them. Dead plants reach grotesquely out of the smelly mass and mosquitoes rise in clouds. Every morning when we walk through the grass, our shoes, pants legs, and dogs get soaked; we keep old towels piled by all the doors to facilitate drying off.

We are still fortunate, because almost every day brings a breeze which makes it harder for the mosquitoes to get a grip, and dries our perspiration

Today I found a letter my grandmother wrote me on April 9, 1962. I must have been asking her about my biological father; her letter testifies to his behavior. *"Darling Please don't worra a bout what you mentioned and believe all your Mother told you a bout a certain party she shurley did hir best to make a go of it thats what that unspeakable indulgence can do to peopel so put it out of your minds and just be thankful you never grew up in sutch environment."*

Yes, Grandmother, I am grateful to my mother for getting me out of that environment. Interesting that she could spell that word, considering some of her other spellings.

July 12. The mystery of my mother's first marriage, to her high school sweetheart Robert Osborn, deepened today when my mother's brother told me that the reason she divorced him was because she'd become convinced he was of mixed ethnicity; she didn't want to bear children with "tainted blood." I recall her showing me a picture of his mother and saying, "See, doesn't she look as if she's black?" The woman looked no more black than I do.

I tried to ignore my mother's racism, but she often remarked on skin color and subscribed to my father's belief that the smart races lived in the northern hemisphere and were white, while the southern races had darker skin and lower intelligence. Surely, with all the reading my parents did, they realized how thoroughly those ideas have been disproved.

July 13. This morning I weeded around the lilac, buffaloberry, plum, and chokecherry bushes I planted north of the retreat house, in the gaps between the pine trees my father planted. Will there be a third generation to plant trees or bushes here?

We've discovered a new-to-us plant on this hillside where we've walked three times daily since 2008 (a couple of thousand walks) and I walked thousands of times before that. Six to eighteen inches tall, it has a five-petal bloom in lavender-pink. The Clasping Venus's Looking Glass belongs to the *Campanulaceae* family and is an annual native to North and South America. The origin of the name, says one source, "is obscure." Some say the seeds are so shiny they are mirror-like, or that the flower resembles the oval shape of an old-fashioned looking glass with a handle.

July 14. Peacefully, I have become a year older. Breakfast was my homemade granola, with a side of the lemon pound cake I made for my birthday.

Today I looked at the autograph book inscribed by my mother, *"To my dear daughter on her 11th birthday, July 14, 1954."* In front, I copied "Horse Sense" by Anonymous:

> "A horse can't pull while kicking.
> This fact I merely mention.
> And he can't kick while pulling.
> Which is my chief contention."

The verse came from my new father, who recited the poem while explaining to me that whining wasn't getting any work done. Complaints could land me in the garden with a hoe.

The pages of the book are liberally sprinkled with names I had forgotten: I wish I'd written about each of them, but I thought I'd remember them forever.

For our birthday dinner, Jerry grilled steaks and I made lime ice cream after a day of doing what we both enjoyed. My gifts were things I needed: good socks, pens, books, a walking stick, and a vest with 21 pockets for our hikes.

July 15. Every morning when I let the dogs out, I try to get through my email quickly so that when I go to my study to work later in the morning I can go straight to writing. When this works, I can ignore the Internet until late in the afternoon. Sometimes I drop in to answer one question and surface an hour later having answered emails that have no importance at all. Infuriating, but my own fault. I remember with chagrin how, before I got email, I scoffed at a woman who said she wasted her writing time answering email; today I apologized to her posthumously.

July 16. Looking through Mother's albums and dozens of loose photos taken during her high school years, the early 1920s, I finally realize she was creating herself.

In one photo she stands like a model, legs akimbo, wearing a silky flapper dress and looking flirtatiously over one shoulder, lips parted like an aspiring starlet, eyes half-closed. In another, wearing a fur coat of my grandmother's, she stands with her legs wide apart, her head thrown back and her hands clenched on her hips. Here she prances with her hands in the pockets of her knickers with a sporty hat tipped flirtatiously over her eyes. She is always looking at the camera; there are no photographs of her laughing uncontrollably or eating.

Every single photograph of my mother is posed.

Yes. I recall once seeing a picture of her taken when she was in her seventies: rushing toward the tour bus that had been waiting for her for fifteen minutes while she made one of her frequent trips to the bathroom. Here she comes, apologizing, rushing through a crowd of folks. I think the photographer meant to remind her of her tardiness rather than to memorialize the moment, but it stands in contrast to all the other photographs and I cannot find it now. Surely she destroyed it.

Of course. How much of her life was a pose? She created a picture of perfection that matched her teenage aspirations to be a movie star, a vamp, a beautiful, beloved woman always prepared for the camera's lens.

She dressed for every photo, chose the setting, directed the photographer, and destroyed the ones she didn't like. Here she pets a dog, looking soulful. She poses with girlfriends, with boys, and with her mother. Several album pages are filled with pictures of young men, "George," "Francis," "Lawrence," "Russell," and all of them posture as well: wearing suits, ties, in front of cars, behind the wheel, leaning against trees, and looking as if they were modeling menswear. Her journal mentions Stella's camera; perhaps she and her friend Stella were acting out their fantasies.

Her life in Pittsburgh must have lasted from her first marriage in 1931 to her second in 1938, but I found few pictures of it. None of them show her with Robert Osborn, though in one or two she is with her half-brother George in his Army uniform. Another photo or two, taken by a street photographer, shows her with her second husband, my father Paul Bovard, his fedora rakishly tilted, both of them smiling in a distant way. In another photo she and Paul are lying on a lawn with another couple, all wearing bathing suits and holding beer bottles.

Under the stairs I find the collage I made for Mother when she moved to the nursing home. I chose pictures from her whole life, thinking to show the nursing home staff something of her personality before she was bent by osteoporosis and her mind was clouded by years. Only when I was putting it together, after she'd left her home and I had access to the papers and photos she'd left behind, did I discover the existence of her first husband.

In one picture she wears a white pants suit and jacket and stands beside a palm tree; in another she lounges on a beach in a white bathing suit. "Those were from my honeymoon with Paul," she said mistily when she saw them. There's a portrait of her the day she married John Hasselstrom and another on the porch of the ranch house holding several kittens.

Finally I look at one that has always puzzled me, a large tinted photograph in which she looks as if her heart is breaking. When she saw it, she said, "Where did you get that?" and refused to explain. I always believed she had just lost, or perhaps aborted, a child, but I have no evidence for that belief. Now I wonder if it was taken after her first divorce. Photographs are supposed to tell the truth, but which truth? Much depends on what we know when we look at a picture.

Only the photos that survive can still speak; the ashes of burned photographs, or of a journal, or letters, might contain a completely different narrative.

July 17. A high of 101 degrees.

A few days after Mother died, I opened a plastic locket she kept in her bedside table. It contained a snapshot of John Hasselstrom in the 1980s and a photo of me at age three. I find this choice significant: she treasured the picture of him as he was in the middle of their lives together. He hated being photographed, so it was surely one of the best pictures ever taken of him. Meanwhile, though I often sent her pictures of myself, she chose to keep close a picture of me as a child. The implication is clear.

Finally I admit what I believe to be the truth: she always took photographs, wrote diaries and letters, and she kept innumerable souvenirs from every period of her life. But she must have destroyed most of the evidence of her first marriage. I found a few of Robert Osborn's letters, the announcement of their wedding, and a few dance cards on which his name appears. I found no photographs, not even of their wedding; no marriage certificate, no documents with her name as Mildred Osborn.

I'm struggling to understand this. In 1973, during her period of demented behavior, she said she never should have left Bobby Osborn. Surely if she always felt that way, she would have cherished souvenirs of that marriage.

I, too, made a mistaken first marriage, but I can't imagine destroying every record of it, because during those seven years my life changed dramatically. Many of the changes were painful, but I can't erase them. What if she had been required to prove that her marriage to my biological father was legal?

Part of my confusion is because I have now read some of her first husband's letters, so urbane, so kind, so utterly centered on his caring for her. I understand that one might not return love in the same fashion it is given, but surely that would not be cause for destroying all evidence of it.

A mystery, and one that I am unlikely now to solve.

July 18. Some days when I drive to the post office in Hermosa, I imagine turning right and finding Harold in his house. Or turning left and pulling into my aunt Anne's yard as she comes to the door with a smile on her face asking if I want coffee. Today I was trying to remember who lived at a particular place and thought that if I could drive around the neighborhood with my dad one more time, listening to his stories about who and where they lived, this time I'd write them down. Now I wish I hadn't been so damn efficient in recycling those old telephone books, which would answer many of my questions about who was where.

What have we lost when postmasters in these small towns are assigned by seniority rather than by living in the community? My local postmaster called me one day before Christmas to say that I should pick up the mail rather than Jerry, because a package she assumed I'd ordered as a gift for him was labeled with what it contained. That's the kind of service that is growing rare.

July 19. Today I saw a robin make a dozen trips carrying mulch from the raised beds to the cedar tree where the snake killed the nestlings: building a new nest. Then I went back to digging in my mother's past.

I'm stunned at how little I know or can prove about my biological father, whom I never saw after I was four years old. I remember him mostly as a voice yelling while my mother broke his whiskey bottles.

Their marriage license shows that R. Paul Bovard married Florence Baker in Morgantown, West Virginia, on April 16, 1938. Born Florence Mildred Baker, my mother was adopted by her stepfather, Walt Hey, and became Florence Mildred Hey. She never, to my knowledge, used the name "Florence;" I've never found it in her handwriting.

If Mother was divorced from Robert Osborn, wouldn't she have been Mildred Osborn? Or if she had asked the court for return of her maiden name upon her divorce, she'd have been F. Mildred Hey. Why did she use her birth name in marrying Paul? Was the marriage legal if she lied about her name?

The amazing world of the Internet helps me reach officials in Allegheny County, Pennsylvania, who tell me that her divorce from Robert Osborn was Case 949, filed in October of 1937; divorce was granted February 28, 1938. My mother was the defendant, that is, her husband requested a divorce from her! The case file has been lost so I don't know the grounds for the divorce.

My mother told me again and again that Paul only wanted a baby to keep from being drafted. Yet he'd enlisted for three years' duty as a private in the Air Enlisted Reserve Corps in 1929 and was honorably discharged January 31, 1935. I have his pilot's log and aerial photographs of Langley Field where he apparently learned to fly; the log is signed by Charles Lindbergh.

He soloed on May 2, 1942, at Cunningham Airport in Houston, Texas, where my mother was living. He had already been in the military by the time I was born, so surely he wasn't afraid to serve. And wouldn't his previous service have kept him from the draft?

When I found this information, Mother was in the nursing home and insisted she didn't know what I was talking about. If this was the only lie she ever told me, I could pass it off as confusion.

On April 22, 1943, three months before my birth, my mother wrote to her *"Honey Bunny"* that *"the colored men washed my windows."* They charged $1.50 but she gave them $1.75. She'd started spring cleaning and said that when he came back the apartment would be so clean he'd think he was in the wrong one, *"but you'd better not be!"* She signed herself as his loving wife *"Milly + ?"* and wrote in parenthesis that she was *"taking good care of it."*

Writing back to her, he sent love to her and *"junior"* adding, *"let's never call him that."* After I was born, he called me "Shorty."

Their divorce decree is dated May 23, 1947 and says, that the "allegations contained in Plaintiff's petition are fully supported by the evidence," but does not state the evidence my mother offered. She was awarded support payments of $75 per month, payable on first and fifteenth days in payments of $37.50 each until I was sixteen. Their community property was 45 US Government bonds, equally divided, but no amount was stated. Custody of me went to Mother, with Paul given visitation rights provided the visitation did not interfere with my welfare. I never saw him again.

Mother said we never received a penny of the support payments, though he occasionally wrote letters and sent gifts. When I was in college, he found my phone number and called once in a while, always drunk. The night I reached home after my first marriage, he called to ask if I "had" to get married. I hung up and began sending my letters to him through Mother, so he wouldn't know where I lived. My next contact with him was when the San Diego Coroner's Office called to ask my permission to do an autopsy because I was his next of kin. He had been found dead in a fleabag hotel with an empty bottle of whiskey, an empty bottle of barbiturates, and only a few dollars in his wallet.

What if I did not agree, I asked?

A dry official voice informed me that his remains would be kept in the morgue until I agreed.

Why was it necessary to do an autopsy, I asked. The cause of death seemed pretty obvious, but of course I gave permission. When I called to tell my mother, she started to cry and sing, "Sweet Leilani" on the telephone. Paul had wanted to name me Leilani, she said. My biological father was cremated, his ashes buried with those of his sister somewhere in Pennsylvania. I've had no contact with the family for years.

July 20. I woke up giggling from a nightmare in which I was in high school and my name was Leilani Hasselstrom.

We finally took my birthday hike today, along a trail in the Black Hills lined with flowers: bee balm, clover, sunflowers, asters, yarrow, and goldenrod, all attended by swallowtail butterflies. A flock of young turkeys scurried through an aspen grove following the hen.

July 21. The sunset light turns the grass gold against a blue-black sky on the south: another thunderstorm. Bolts of jagged white light slam into the ground but the sound is distant. At dusk yesterday we watched hummingbird moths dipping into the yellow trumpet blooms of evening primrose. Today I found one of the moths dead; grasshoppers had eaten off its wings.

Late this afternoon, we escaped the heat with a drive in the Black Hills. On a gravel road, I saw wild pink roses like those that lined the road to my grandmother Cora's house, where my parents often went on Sunday. I sat in the back seat of the 1952 Chevy trying not to wrinkle the dress, always pink, my mother made me wear. Sometimes I smuggled a book along; mostly I looked out the car window. No matter how hot it was, my mother kept her window rolled up so the wind didn't muss her hair; rarely we could persuade her to open the wing window to let in a little air. No air conditioning, of course.

Writing this, it occurs to me that these days car windows rarely "roll" down, and anyone under fifty may not know what a "wing window" is. I miss them still; they allowed regulation of the air in the way that a single pane window and vents on the dashboard do not.

But I digress; therefore, I am a writer. Petals on the low pink roses ruffled as we swept past, the breeze lifting dust. We saw the roses only in the foothills and deeper valleys of the Black Hills. Because I only saw them from the passing car, I never smelled them, and we never stopped to pick them. They looked fragile but survived the gravel roads and buffeting from traffic.

I took pictures and told Jerry about those trips to Grandmother's house while I scribbled reminiscent notes into my journal. I've been so busy with volunteer projects, retreats, gardening and harvesting that I hadn't written anything for days.

As soon as I got home, I looked up "wild pink roses" and discovered they are the only North American native rose, *Rosa blanda*, usually called "Prairie Rose."

Prairie Rose! A new dimension. Prairie Rose Henderson is my favorite among the old-time cowgirls, for her wide smile both before and after she lost her front teeth in a rodeo accident. (I refuse to digress again as she's become a popular figure.)

By this time I admitted that I was writing—to use its ugly modern name—a "blog post." I don't know where else this topic might lead, but the photos and notes have served their purpose. I am no longer frustrated by not writing. I have written something that's intriguing enough to lead to other writing. That's what writers do.

July 22. At noon, walking the hillside above Lake Linda, we noticed that the duck families swim for the center when they see us. First one hen, followed by six ducklings, then another hen with nine, then one with seven slightly larger babies. Soon we'll notice that these families are dwindling, fallen prey to herons, owls and snapping turtles.

Rearranging files in the basement, partly as a result of this journal exploration, I've found things to throw out, like high school research papers. I'm taking dozens of books I'll never read again to the local library to shelve or sell, and hauling financial files to the basement of the retreat house; I suppose there may be no real statute of limitations of what the IRS might someday want to know.

I keep thinking about Mother and the dissension that surrounded us. I have sometimes wondered, though not for long because it is a waste of time, who else I might have become. I became a rancher because my mother chose to marry a rancher, but some of her suitors lavished attention and expensive dolls on me; she had choices.

Would I have become obsessed with money if she'd married that banker who courted her? She spoke of John Hasselstrom as a wise investment in her future and mine after she discovered his financial status. She made it clear she was sacrificing her career as a legal secretary and the culture found in town for security. Since she didn't like to get her hands or her shoes dirty, she fought ferociously with my father to keep me inside the house learning to clean and cook. She tried to train me to be a wife, preferably of a doctor or a lawyer. She didn't want me to marry a writer or a rancher, let alone *be* both of those things; until the end of her life she often asked if I gotten a "real job" yet. Yet she loved to "drive out past the ranch" when she lived in the nursing home.

She never asked questions about how I was juggling my own life and managing the mess my father had created during the last few years he lived, or how I managed after divorcing my first husband or after George's death. My father left the ranch to Mother; he mentioned me only to say that he intentionally "omitted to provide for" me and his omission was "not an oversight." After my mother's attorney told her how much the ranch was worth, she said, "I'm rich!" and began planning to sell it. Instead, I sold my cows to make a partial purchase payment, and her attorney persuaded her to bypass the rest of the ranch to me. She never realized that when I accepted a speaking engagement, I was raising money for fence repair and building maintenance. My tough uncle Harold cried with me when I told him I had to sell my cows, but to Mother, cows were only money.

I respect my mother for many reasons. Her culture taught her to be a wife who stayed married to one man lifelong. Yet after two disappointments in marriage, she supported herself and her child in an era when being a single mother was more difficult and much more unusual than it is today. Her marriage to John Hasselstrom was the practical choice of a woman 43 years old in 1952, when being a housewife was a woman's ultimate achievement. She said laughingly and often that she didn't want to know the cows' names, just how much they brought at the sale ring.

The ranch was not a way of life for her but a means to an end. She never joined the community my father and I formed with the landscape and animals. Instead, she looked over her shoulder, longing for the city life she loved when she worked. Perhaps, though, Mother did not become part of our family society because my father and I shut her out. We worked outside most of every day and our conversations in the evening were about things she had not witnessed. During the early years, she tried to help us outside. After several spectacular failures—getting lost, wrecking the pickup—she returned to her housework. She worked hard to make our lives comfortable, and my father talked to her about ranch work, but I don't believe she listened.

July 23. In my diary for July 23 to 26, 1954, I wrote one page per day: "HAYED ALL THIS WEEK." That meant I was driving my little 420 John Deere tractor hitched to the dump hay rake, reaching back to pull the rope that dumped the hay each time the roll of alfalfa grew large enough. When I complained that my right arm would be twice as long as the other from this exercise, my father said, "Switch arms."

I visited with my mother's half-brother this week. He believes she left Paul in Pittsburgh and fled to Houston, Texas, where her mother's sister lived, to get away from him. He followed and must have persuaded her to give him another chance. I remember leaving Houston; she carried me onto a train at night: I can still see the string of lighted windows behind her. My uncle told me that we rode the train to Wheatland, Wyoming, where we stayed briefly with relatives. Then she left me with Grandmother Cora at the Hey ranch near Edgemont while she found a job and a home in Rapid City. My grandmother came to live with us for a while, but I remember walking to second grade alone or with the boy next door.

Then until I was nine years old, Mother was single and working. I remember walking with her to wash our clothes at a laundromat. And walking to the top of a hill to fly kites in the spring and ice-skating on a pond beside the creek in winter. She wore her big fur coat; I wore mittens my grandmother had knitted, tied together with string that threaded through my coat sleeves. A photograph shows me and my skinned knees standing with my grandmother on the sidewalk where I roller-skated in summer.

July 24. Last night I dreamed about a photograph I've seen, of me sitting in a frilly dress on the sidewalk in front of our apartment in Houston, the concrete warm against my bottom. What doesn't show in the picture is the memory. Two sidewalks lead up to that apartment house, with a long rectangle of dark earth between them, thick with strawberry plants. In my dream I remembered the old man kneeling on the concrete to weed the strawberries. He wore a narrow-brim straw hat, a long-sleeved shirt with buttons down the front, tan pants worn beige from washing, and heavy shoes. His hair was white and he had bushy eyebrows. He asked me if I'd like to have some strawberries. Of course I said yes.

Where was Mother? Surely she was nearby.

"Go and get your mother's sugar bowl and I'll show you how to eat strawberries."

The sugar bowl was white with flowers painted on a glossy yellow rim. I brought it to him and set it carefully on the concrete sidewalk. He picked a half-dozen strawberries and sat beside me. We each took a strawberry. He bit the end off his, then dipped it in the sugar bowl. I did the same. Dip, bite, dip, bite.

I can still see how the red drops of strawberry juice collected sugar and gathered in the bottom of the depression we made in the sugar. Juice and sugar dripped on my dress, on my hands and arms. We grinned at each other and kept dipping and eating. I'm sure that the man knew my mother would be furious, and surely she was, but I recall none of that.

In my dream, as in my memory, the scene does not end; we are still sitting there, wrapped in warm Texas sun, dipping and eating and smiling. At that time, I remember mostly turmoil in my house where my mother was always screaming at my father if he was home. Surely the old man and those strawberries was one of the few good memories I have.

July 25. Each morning I work with the writing of the woman at the retreat. She has been writing about her mother, and we've had some discussions that penetrated deeply into our beliefs, our fears, and our ideas about our mothers.

Today's the day, I promised myself this morning, just as I did yesterday and the day before. Yes, today's the day I write my website's home page message for Lammas, the ancient Celtic festival that occurs on August 1, celebrating the end of summer with harvest, and the beginning of the year's slide into winter.

Yesterday while not coming up with any ideas for the home page message I checked on the garden and did some filing in the retreat house, moved my refrigerator out of its niche and cleaned under it, vacuumed its coils, and washed the outside. Then I vacuumed and dusted the house and aired some rugs. After lunch I finished up the plans for a workshop I'm giving tomorrow, including making a decision about what to wear. I used my pocket watch to time myself and took some photos. None of those activities produced an idea for the home page message.

By 9 A.M. today, I'd written a few thoughts—not about Lammas—in my journal. After breakfast I tidied the kitchen, walked the dogs, planted wildflower seeds, and bathed the dogs while deciding what to fix for lunch. Cleaned the washer, dryer, and utility sink inside and out, and dusted and scrubbed the basement bathroom.

Most of my housework gets done while I'm avoiding writing.

I love writing; it has provided some of my greatest joys, especially that moment when I've finally shuffled the words enough to find the perfect phrase.

But it's also provided hours of staring into space, unable to think what words need to come next. So my subconscious and sneaky brain can find all kinds of really good logical ways to avoid it.

As soon as I sat down at the computer, I needed to change the location of the water on the garden. I rode the four-wheeler down and sat on it with my garden plan, comparing that glorious vision I created while planting seeds this spring to the few plants that the voracious grasshoppers did not eat. We've had more hoppers here this year than I've ever seen. Neighbors who drive through have been shocked and said they have fewer.

Around my house, I've noticed which of the Natives and the Nasties have survived the hopper onslaught. The grasshoppers have avoided native plants like buffalo grass, sideoats grama, and gumweed. The Nasties—introduced plants like brome, alfalfa, mullein, and clover—have been stripped of their leaves and then their stems. Unfortunately, non-natives I like—columbine, peony, chamomile, arugula, marjoram, thyme and dill—were decimated. Apparently even grasshoppers don't eat creeping jenny.

While I looked over the garden, I kept thinking of Lammas. How could I write about harvest with no produce? How could I write about harvest when my summer has been seriously unpoetic, with a variety of activities and responsibilities disrupting my writing?

Today, walking among the plants, I noticed that only a few hoppers leapt away from me, instead of the moving blanket of three weeks ago. Pulling bristly foxtail from the leek row and stuffing it into the composter, I saw that the tomatoes are strong and blooming. The pumpkin vines sprawl and blossom, leaves shivering as entire rabbit families lounge in their shade. The kale and turnips are getting taller.

Back at the house, I saw that in my kitchen garden leafless tomato plants are bringing forth yellow Taxi tomatoes and Early Girls! A couple of pots of basil and parsley so big I couldn't move them are putting out new leaves. The garden is working hard to recover from the many failures of the summer. Maybe I can give thanks for some growth and will have a subject for the harvest message.

Inside, sitting with my fingers on the keyboard, I saw a bird that sent me scrambling to my bird book, a male orchard oriole that landed in the raised tomato bed. He clung to the tomato cage, tilting his head this way and that and then hopped. Hopped. Hopped again and snatched a grasshopper. Gobbled it and hopped some more, following hoppers as they tried to evade him. Yes! In writing, too, we sometimes have to chase ideas, be persistent, leap and snap and gobble.

Later, I stepped outside into a maelstrom of clucking and fluttering: two grown grouse and eleven teenagers were all scrambling around the dogs' small pen, eating grasshoppers, and chattering to one another. I went back to the computer.

My friends kindly say I accomplish a lot. Much of what I do, however, is pure avoidance of the job I both love and find most frustrating: writing. I have had two big writing projects simmering all summer, getting to them only in short bursts.

So, in the spirit of this season of harvest as the end of summer, I faced my failures: I have not yet finished the draft of the next book I hope to see in print. On August 1, I will follow Celtic tradition and write that failure on a piece of paper and burn it symbolically, carefully, in a candle flame in my study. I'm too nervous about prairie fire to create a traditional fire outside.

Once I have become fully aware of my regrets, I will let them go. If I spend time brooding on what I failed to accomplish this summer, if I attempt to figure out why I did not do all that I wanted to do, I will be wasting time during which I could be writing.

Farewell to all the plants and seeds I consigned to the ground that did not grow and bloom. While walking the dogs, I pull seed heads and trample them into the ground around the house and yard, thus planting them with hope for next year.

I've dug the potato crop, and we will eat every bit of it (three potatoes) one day soon, trying not to think about last year's crop, which supplied us with potatoes from September through May.

At the end of my calendar, where I list the things I have accomplished,

I wrote a workshop proposal to the National Cowboy Poetry Gathering in Elko, Nevada, which was rejected, but I've revised the workshop to use in another context.

I've written four home page messages, one each for February's Brigid, the Vernal Equinox of March, April's Beltane, and the June Summer Solstice, a total of almost 9,000 words. I wrote the introduction to a book by a writer who has attended retreats at Windbreak House, as well as cover comments and reviews of other books. Observations about meat, grouse, natural predators, rabbits, organ meats, snakes and other prairie critters all furnished subjects for blogs. I responded at length to questions sent by a college class reading my book *No Place Like Home*. I wrote two essays published in *Orion* magazine. Later, National Public Radio's "Living on Earth" asked me to read them for on-air publication. A request for free writing advice turned into a lengthy blog on why I won't provide it. On paper, I reflected on the fact that I am called a "nature writer," later submitting the essay to the International League of Conservation Writers, which published it online. Besides all this, I wrote often, either by email or by mail, to several friends.

Compiling this list amazes me. Though I was determined not to regret what I have not written or done, I hadn't fully realized how very much I have accomplished so far this year. Truly, my writing harvest has been generous.

And now I realize that I have the draft of my Lammas message for my home page.

July 26. Each afternoon a thunderstorms rises black and threatening. Each storm rattles the windows, throws deck furniture around, and sneezes a few drops of rain.

In the evening we sit under the deck, chatting, drinking, and watching the tomatoes. When one of us spots a hornworm, we snap it off and squash it, gardening while relaxing.

What I remember most clearly about my mother's house in Rapid City is the kitchen, with its tiny enamel-topped table and my high chair with thick layers of paint. My single bed was behind the door in Mother's bedroom; Grandmother slept on a foldout couch. The house is gone; the firm which insures my ranch property stands on the site.

Every Christmas, Mother's boss at the law firm came striding up the sidewalk, his hat pushed back to show his bushy eyebrows, long coat flapping around his ankles, and a big smile on his face. Under one arm would be a box containing an exquisite doll. The one I remember most clearly was supposed to represent Martha Washington, with a long silk skirt, pearls, white hair piled on her head. Mother wouldn't let me play with them; she wrapped them in paper and kept them in boxes. When we moved to the ranch, she put them on the top shelf of my closet. I found them there after she died and took them to the Salvation Army where I hope they were bought for little girls who lovingly played with them until they fell apart.

July 27. When we lived in town, there were many things for a woman and child to do within walking distance of our house; we were practically downtown, close to the library, movie theaters, and restaurants. We walked everywhere.

On the ranch, the nearest neighbor was a couple of miles away, across the highway, so my father taught Mother to drive. I remember her slamming doors after they came in from driving around the pasture. In the early years she often drove to one of the neighbors to visit or to trade eggs for cream. She joined Ladies' Aid and the choir so she could "get out." She commented that the neighbor women didn't have her interests: they didn't read as much, and they worked outside with their husbands.

My new father drove for most of our shopping trips; they'd start on one side of town and work their way to the other, getting groceries last so the frozen food wouldn't thaw on the way home.

Now that I think of it, shopping with my impatient father must have been a trial. I often cringe when I see couples shopping together, the man questioning every decision the woman makes. The woman is distracted, trying to find bargains while explaining why she gets this brand instead of that. I hope sometimes my father dropped her off and did other errands.

Mother was once Queen of the May but after she married a rancher she was no longer in control. My father's dawn-to-dark work schedule controlled what we did on the ranch. As soon as I got old enough to question her judgments, she didn't control me either. She must have felt adrift.

July 28. We decided on an escape, loaded up the dogs, and took a great drive to Pringle and one of my favorite spots, Point of Rocks. Part of this land was once for sale and George wanted to buy it, to build a house among the limestone boulders and grow grape vines on the south side. We knew it was purely fantasy, yet some part of me saw us nestled into that hidden spot as Jerry and I drove past. I haven't been there since George and I dreamed that dream.

Then we drove up and around and around on steep gravel roads past Hazelrodt campground, a place I'd never been before, and emerged in Custer. Nice walk with the dogs, relaxed observations, and conversations. Stopped for ice cream and pie at the Purple Pie Place, listened to the tourists being amazed at the quality of the desserts and smiled, thinking "We live here!"

Lately I've found myself turning back to a scrapbook I started when I was eleven or twelve years old. At first, I was disappointed in my creativity. Why did I select these items? The book begins with a picture of a little blonde girl sitting on a stone branch under a tree, hands clasped, and face smiling in delight. My cynical adult self calls it sentimental; Mother no doubt pasted it in the album because she had a big picture of me in nearly the same pose, with blonde ringlets down my back. This is how she saw me and how she wanted me to record my life. On the next page is a picture of a baby with a towel wrapped around its head; I have never liked babies, but my mother never gave up trying to grab any baby in sight.

Still, when I was eleven years old, I cared enough about some of these things to fix them on these pages with flour and water paste my mother taught me to make. So far this summer, I've been inspired by the scrapbook to write about the janitor in the Hermosa school and the watch he kept in his vest; about our Rapid City neighbors when I was five years old, Mrs. Bradley and Mrs. Melaven; about how an old man taught me the proper appreciation of strawberries when I was three years old; about outhouses and the courthouse in Rapid City; about my musical career, and about how my father John pronounced "pizza" (with a short vowel and lots of emphasis on the z's.) I have no idea what, if anything, all that writing will become. I don't worry about it because I am writing. I feel better already than I did yesterday; I am writing.

Thinking about all that my father taught me, I've naturally been working on a poem:

LEARNING ABOUT GATES

Always shut the gate behind you,
he taught me first when I was ten.

He showed me how to squeeze hard
and yank the wire loop up. Before
my muscles grew, he let me use
the wire stretcher on the tight ones.
When we moved cows to summer pasture,
he'd drive the pickup ahead, open each gate,
count the cattle through while I followed
on my horse. He'd shut that gate, idle
the truck past, being sure each cow's calf
was next to her; open the next gate.
He didn't go to church.
He put his faith in keeping fences tight
and four wires high; in shutting gates
and paying bills. He believed in buffalo grass
and rain in June. He didn't trust anyone
who couldn't look him in the eye or shake
his hand. He said "My shadow on this place
is worth forty dollars a day."
At his funeral, the minister said
God planted a garden over east in Eden,
and created ranching when He made a woman.
My neighbor Margaret told me,
"He's gone ahead to open the gates."

Gate Keeper, whether your gate's gold
or barbed wire, you can trust my dad
to count them through.

July 29. When I open my journal in the morning, the pages feel damp and sticky with the humidity. Jerry mowed the entire hillside and dug up more of the rocks that damage the mower. He's adding them to the cairn George built at the hill's highest point and filling in the holes so we don't break an ankle.

Another memory of my mother. When she got a little black dog, she named him a derogatory term many black animals were given at that time. When I brought my stepchildren home for the first time, she was standing on the porch when they asked what the dog's name was.

I looked my mother in the eye and said, "Blackie."

She laughed and said no, it wasn't, and told them the right name and laughed.

"My mother can call the dog that if she wants," I said, "but we do not use that word in our family, and you should never say it in my hearing."

She called the dog loudly and often whenever the children were around.

At sunset, a nighthawk dives from the gray sky toward silver grass; just as it strikes prey and booms, the last sliver of sunlight disappears.

July 30. Jerry often whistles as he makes breakfast in the mornings, and then may whistle snatches of song all day long, a cheery sound I love.

Yet much as I love music, I never whistle, or hum, or sing. Listening to him this morning, I recalled how often my mother sang a few lines and her love of music.

Sometimes the snatch of song that clings to my brain has some significance to me. When I hear "Little Mountain Church," surely one of the most sugary, sentimental hymns ever written, I get tears in my eyes because I can hear my mother singing it. Mother tried desperately to interest me in music before she realized I couldn't carry a tune and didn't have much sense of rhythm. She enrolled me in piano lessons from a woman I remember primarily for her harsh voice, her metronome, and her shellacked silver hair. She loved my hands because I could reach more than an octave, and I learned to play the piano fairly well.

Mother insisted I take high school choir; I didn't last long. As soon as the instructor heard me sing, he asked if I could play the piano. I said I'd joined choir to learn to sing.

He said, "You aren't singing in my choir; you will play the piano."

"Isn't it your job to teach me to sing?" I said.

"Play the piano or get out." I quit the choir.

July 31. We harvested several ears of corn last night; very tasty with grilled burgers.

The thermometer stood in the 90s all day. At noon a thunderstorm crashed; one bolt of lightning struck so close to the cattle I expected to see one dead. At dusk we sat on the deck enjoying a light breeze; even though it was still above 80 degrees, the illusion was of peace and cool since the day was so hot and busy. Lightning flashed in the south and a fragment of rainbow shone in the distance; we had no rain.

Just after full dark a full moon rose, and the coyotes howled high and broken cries from the south ridge.

— August —

August 1. At noon the cattle lie in a half circle along the edge of the gully below the house. My father always said, "You know cattle are doing good when they lay down in the middle of the day." The gully is damp, with cattails growing in the deep parts.

I miss the curlews—big birds who landed with their wings straight up in the air, like confident ballerinas on tiptoes, then slowly drew the long wings down against their sides. Maybe they moved because the previous years' drought dried up the little pockets of water they found in the limestone ridges, but they're gone.

The Sibley Guide to Bird Life & Behavior tells me the Long-billed Curlew, the species we had here, is a threatened species, with low population numbers and limited distribution. Some species of curlew are believed to already be extinct.

I'm suddenly crying as I realize I may never again see that great M shape as a curlew settles onto a nest. Yet I can think of the deaths of my parents and even George without tears. What is the difference? Perhaps it's that my loss of my loved ones is private, whereas millions of people who don't even realize what they are missing will not see the curlews.

Nature is filled with such tradeoffs; but sadly, most people who get a "little acreage in the country" probably remain ignorant of the damages they cause. I can't blame loss of wildlife solely on expanding subdivisions, though. The barn owl and American kestrel are both declining in the northeast, where farming has declined and forest has increased. What ingredient is missing in the habitat that those birds need? How might I bring the curlews back?

Further reading convinces me that the curlew needs precisely what I am already working to preserve here: grasslands uninterrupted by agriculture, by development, or by crowds. Curlews need to be left alone to thrive in grasslands, just as cattle do. I'll hope that the reason I haven't seen them is because, without horses, I'm no longer taking long rides alone into the most distant pastures.

August 2. An eight-inch-tall burrowing owl stands beside its adopted home—a prairie dog hole—and stretches its wings, glaring at us. Unlike most owls, these charmers forage by daylight, mostly on insects. While the burrowing owl is threatened in the Great Plains because of urban sprawl and farming, as well as poisoning of prairie dogs and the plowing of the their burrows, this species is actually thriving on golf courses, especially in desert regions, because they can find water, cover, and insects for food. Not all change is inherently bad, even for the animals.

I used to love seeing those little owls when we rode through the prairie dog town. They stood firm on their mounds, hooting even as the horses nearly stepped on them, but they did not fly away.

At sunset, blue-gray clouds stack high on the eastern horizon. Behind them one rolling billow of white silk rises, shining, slowly turning apricot-colored as the sun sets in the west. Nighthawks wheel in dark arcs. When the clouds have all turned blue, the moon rises like a ripe peach, fading into gold as it climbs above the billow of clouds.

August 3. A band of gray hangs above the horizon on all sides, a circle of smoke from forest fires. The rumble of motorcycles hangs in the air all day long: bikers headed for the Sturgis Motorcycle Rally.

The cattle have been moved to fresh grass from the pastures closest to our house now. Lake Linda has dwindled in the heat and is surrounded by a broad band of cracked mud pocked with deep hoof prints. If we get rain, the grasses can begin to recover. We don't water the grass in the yard, but in this dry spell we've noticed buffalo grass moving into areas where the drought has killed some of the weaker invasive weeds. The barest whiff of moisture—a tenth of an inch or less—greens the buffalo grass at the base.

Today I read back through what I have been writing and was astonished to find that I had told several stories more than once: about the bloody white dress, the whiskey bottles broken in the sink, Mother's lie that the martini and cigarette picture was a joke. I was amazed at the variety of incidents that reminded me of these stories, as if my brain was looking for an excuse to dredge them up. Clearly these events have haunted me. Can I now banish them? Surely this book must be my final word on the subject to myself and others: an exorcism of whatever demons they represent.

Many writers, theologians, and therapists have said that silence protects no one. I was always told to keep silent about trouble, to "keep it in the family," but doing so only encourages more silence, and perhaps more silent suffering. Will blabbing my family secrets encourage someone else to delve into her own history and change her life? I hope these revelations are useful, not merely gossip.

Perhaps I have already performed my own exorcism, by trying to discover the truth in the lies. Perhaps now I can turn away, focus on good memories of my life with my mother.

August 4. At this time of year we suffer from sticker shock. We wear high boots against rattlesnakes, but our socks fill with grass awns anyway; they stick in the dogs' ears, in shirts, gloves, even in underwear. We sit on the deck to pick them out, putting them in a container to burn because if they blow off the deck we will only have succeeded in planting noxious, invasive weeds in a new location.

A friend says if you get bristly foxtail seeds in your socks, you might as well throw the socks away. I can't do that, so I try to remember to wear the same socks every time I go to the garden where the seeds were washed in by flooding. Eventually the nasty seeds seem to find every pair of socks.

August 5. Our chokecherry and buffaloberry bushes show no fruit, though in many drought years fruit has been abundant.

Looking at my 1981 journal, I see that we poured concrete for the north half of the basement floor of this house on this day.

Twenty-nine years ago today George and I left for a rendezvous with the Yellowstone Mountain Men of Cody, Wyoming. After we put up the lodge I wrote how the cover smelled of wood smoke evoking "freedom and relaxation and nostalgia." George had been welcomed as a member by the group right away; they didn't ask me to join until several years later because they thought my feminism might corrupt their wives! I loved rendezvous, and especially with this group at this time. They were all armed and alert, but despite widely varying political views, we could discuss all manner of subjects with lively enjoyment and no bloodshed.

My nostalgia for those times is eased because the rendezvous that I wrote about in my first book, *Windbreak,* probably no longer exist. The last time I went to one, when George was alive, we vowed not to go to a national one again. Too many people who didn't understand the principles had invaded, parading in purchased finery, waving weapons, and bringing illegal drugs to the campfires.

August 6. When I'm working in the basement, the motorcycles rumbling by on the highway sound like thunder so I keep glancing out my window, hoping I'll see rain clouds. Then I remember: no, just bikes headed for Sturgis.

Thinking about water, as I often do on late summer days, reminded me how much effort we've expended on finding it for our cattle. From my deck I can look southwest to an excellent pasture we hardly ever use, because it has one huge flaw: it has almost no natural water.

Years ago, someone built a solid earthen dam nearly thirty feet tall across a gully that is deep and broad but short. Unless a cloud hovers directly over it and pours rain, or the snowfall is heavy, the dam is usually dry. We always checked it early in spring, and if the dam held any water at all, we got the cattle in there as quickly as possible, knowing we'd have to take them out as soon as the dam was empty.

My father considered drilling a well close to the highway and its adjacent power lines so we could run a pump to fill a tank; we both witched for water and found nothing. I've never understood how dowsing works, but some folks can hold a forked willow twig (or other objects, depending on the dowser) and find water. Both my father and I were able to find water at different locations by this method, just not in that pasture.

We often talked of how we might provide water so the pasture would be more useful. When construction crews put in a new underpass, father had them trench in a length of plastic pipe beside it. He talked of connecting it to a water line so he could pump water from my house a half-mile away from the highway, through the underpass, and into a tank. But the costs for electricity and pumps were forbidding, and his cow sense argued against having the only water in one end of the pasture. Cattle would spend hours walking to get a drink.

Other disadvantages argued against putting a tank close to the highway. Passersby in ranch country, perhaps folks who think ranchers just sit around counting our money, sometimes entertain themselves by putting bullet holes in mailboxes, highway signs, and stock tanks.

We considered drilling a well above the dam so the pond would catch any overflow but a flowing well was unlikely and running power from the highway to a pump on the well would be expensive. Father never had much luck with motors; his sheds were full of electric and gasoline motors he'd abandoned when he couldn't get them to work.

I loved to ride the highway pasture. My favorite part was its secret. In a deep ravine I call the Big Hole in the southwest corner, after a wet winter, I could find natural spring just the size of my cupped hands. Both my horse and I often drank there.

At one time that spring may have been larger. I found tiny dams across the gully above and below it, but no trace of a house or foundation. In one year out of ten, after heavy snowfall and spring rains, the spring swells enough in its narrow gully to support ten or fifteen yearling heifers.

Several coyote dens spill fans of gravel on the slopes above the gully. If I approach the rim of the Hole quietly, I usually see antelope and deer grazing the bottom. Owls nest in the big cottonwoods and bald eagles cruise over. I've found the tracks of bobcats and what a game official told me, from photographs, was "either a really *big* bobcat or a small mountain lion."

Whether it was hand sized or larger, the spring is a favorite of the neighborhood wild ones. Several gullies fanned out east and west of it, affording an invisible approach; other wrinkles in the land gave cover for a wary animal. Always, we could spend a half hour reading the tracks. Before the spring was dry, my father would say it was time to begin hauling water. "Leave some for the wild critters," he'd say and smile.

I wonder how those critters are faring now that there's a subdivision across the fence. Whenever I've visited the pasture I've seen evidence that some of those folks illegally ride their four-wheelers in my pasture.

August 7. At dusk, it's still 80 degrees outside so I take a cool shower and put a fan in the window before I plump the pillows and lean against the headboard to read myself to sleep. The windows are wide open so I hear sparrows cheeping in the grass, crickets among the rocks, even mourning doves far off. Later I'll hear coyotes. Do the people in their air-conditioned subdivision houses hear the life of the country they've moved here to find?

I'm still thinking about our efforts to find water in the pasture across the road. Most summers my father and I would drive to the pasture for the first time in spring, idly commenting on the grass, the weather—anything but water—until we reached the flat above the dam.

Then we'd both get quiet, and the tension inside the pickup would rise as we drove to the gully's lip. From the last flat spot, we could see the wall of the dam crossing the gully and a corner of the pond behind it.

"Oh?" he'd say, smiling. "Looks like we might have some water."

We'd climb out of the pickup, and I'd usually trot ahead to the rocky ridge where we looked straight down on the pond. Then I'd be silent, waiting for him to walk up beside me and see for himself.

"Humph," he usually said. "Now you'd think with all that snow we plowed through in the yard this winter, some water would have had the decency to run down into that dam."

"But it was all in the yard," I'd say, laughing with him. "Maybe we ought to shovel it in the pickup every winter and dump it in the dam."

We'd study the little pool of water in silence for a few minutes, then he'd shrug and turn away. That summer, as usual, we'd leave the pasture vacant unless grazing became too sparse elsewhere, or until fall, when we were nearly ready to sell the yearling steers and heifers. Then we'd turn them in and start hauling water every day.

Each afternoon I'd fill the three-hundred-gallon plastic tank in the back of one of the pickups from the hose in my yard, haul the water to a stock tank on the plateau above the empty dam, and dump it. As soon as the pickup growled through the gate, the yearlings would fling their heads up and amble toward the tank. Once, one learned to drink straight from the outlet pipe, gulping and swelling until I thought his belly would explode.

On a hot day, when I hauled water without my father along, I might strip off my jeans and shirt and slide into the tank with my Westie of the time, Frodo. When the cattle arrived, we'd look up to see a ring of curious bovine faces. A brave steer might snuffle the water, and nudge my naked back with his nose or tentatively lick me with a tongue raspy as heavy sandpaper. I learned to love the chore, the last one of each day. The sun set as the final gallon or two of water glubbed out of the hose. Later we'd sit on the deck with sandwiches and a glass of ice tea or gin and tonic.

About the fourth summer of drought, however, just when I'd perfected my system of hoses and the best methods of filling and emptying the tank, my father discovered solar power.

I think one hot day when he'd had to haul the water, he went inside to work on his accounts in the shade and noticed how much money he'd tucked away. I imagine he could think of nothing he wanted except more water. On our trips to the pastures, he began eyeing likely spots for dams or wells. He was thorough; he carried a willow switch and watched the notches between hills for swamp grass indicating a spring that might be tapped, then he'd dowse to see if the pull was strong. We'd talk over sites one of us had spotted in each pasture and debate whether a well would be better than a dam.

"Or there might be enough water for a dugout," he'd say, gazing at the sky, pulling at the loose skin on his throat.

"Don't you need more water flow for a good dugout?" Successful dugouts are simply sunken tanks, dug into an area where the water level is high enough to allow water to flow in from several sides, or dug into a natural spring.

"But once it's dug, you don't need a pump to raise it." On the other hand, dugouts become less useful over the years, as cattle walking around them push excavated dirt back in the hole. A dugout can become a muddy trap for cattle.

Eventually, he hired someone to drill a well in the highway pasture and put a solar pump on it. Either the amount of water was inadequate or the pump wasn't powerful enough; the system simply did not work. We went back to hauling water.

All that meditation on the waterless pasture was not just idle remembering because in spite of this history, I've contracted for a well to be drilled in the same pasture. I spent yesterday driving the pasture with Rick and the well driller who bought the rig my father had hired on other occasions. We'll drill in the spot the driller believes will be best. Unfortunately, it's close to the highway, so some disadvantages will apply. Still, if we get water, the pasture will become more useful to us. Reading in my journals all that we went through trying to get water helped me make up my mind. I could picture my father smiling. "Don't spend any money," he'd often said, but he'd approve spending it for water.

August 8. My mother's journal reminds me that my grandmother Cora died on this day in 1980. Cora was the only grandmother I knew as an adult and I miss her still, in spite of the fact that these days I see her face when I look in a mirror. My expression is not as sweet.

My other grandmother, my biological father's mother, exists for me only in a story or two. She was a tiny woman and someone once told me that she fell in the bathtub when she was in her nineties and "broke a chunk out of the porcelain tub." I still find that hard to believe, but every time I was stubborn about something my mother blamed in on that Bovard hardheadedness.

I met John Hasselstrom's mother Ida only once or twice; she was in a wheelchair and not interested in me, but I cherish a photo of her standing in front of the old house. She's wearing her best going-to-town coat, her left hand fisted in the pocket, her right grasping a crutch that wrinkles her coat under the arm and makes her head tilt to that side. Her broad-brimmed hat sports a grosgrain bow and is set square on her head, shading her face so it's impossible to see her expression. Her body language speaks loudly: "Hurry up and take that picture so we can get on with it." I wish I'd known her.

All these women are gone, but they still provide me with examples of how to live, nurturing from the grave or from fading photographs.

August 9. I woke up this morning remembering "Linda's Folly" and laughed out loud. When I was sixteen, my father put me in charge of providing a new water source and the project did not end as I had hoped.

I'd suggested the spot, because a neighbor had a huge flowing well less than a mile north of it. Father hired a dragline crew to arrive on a particular day, then discovered my mother had made appointments in town for the same day. He told me to make the judgment about whether the amount of water justified the digging.

"Look for a good strong flow," he said. "You'll usually see clear water in the muddy stuff that seeps in while they're digging."

"How much is a good flow?"

"You'll kinda have to decide that. Think about how long it would take to fill a good-sized tank."

The crew was two younger men and one angry old one. He introduced himself and said, "These are m'boys," gesturing toward two hulking men with extremely short foreheads. Both of them peered at me from under their hat brims, spit into the dirt at their feet and grunted. When they smiled, their mouths were filled with brown, jagged teeth.

My father said, "Follow my daughter to the pasture; she'll show you where to dig and decide whether there's enough water for a dugout." He raised his hand to them, gave me a long look and got in the car; Mother was already in the passenger seat, fanning herself with her grocery list.

I drove my pickup and the men followed with the truck hauling the dragline. They parked above the gully and walked with me down into the bottom where they followed as I paced off the outlines of a good-sized dugout.

Then I perched on a rock above the spot to watch. The old man drove the dragline while the other two stood with their hands in their pockets, occasionally looking back at me from under their filthy hats and muttering to each other.

The engine growled as the bucket of the dragline scooped up rocks, grass, and dirt, which the man dumped on the high side of the gully, close to where he was working. I knew that was wrong; the dirt would be knocked back into the trench as soon as cattle came to water if he didn't move it, but I didn't want to interrupt. The two sons moved rocks away from the machine's metal tracks as it backed up, digging slowly.

By the time the machine had dug the test trench fifteen feet long and a couple of feet deep, it was slowly filling with muddy water. Good. Now he'd realize he had to pile soil farther from the edges. I dreamily pictured antelope slipping up the canyon to drink. I could hardly wait to taste the water in my dugout before any cows walked in it.

The old man scratched around a while with the digger, deepening the trench but barely widening it. After an hour or two, the machine stopped, and the old man got down to talk to the boys. Water bubbled around their ankles, and they all glowered up at me, so I stood up to go down, thinking they wanted to talk about the work.

Then the old man turned off the machine and gestured to one of the boys, who turned toward the truck. The old man started up the slope toward me, puffing.

I met him halfway.

"Not enough water," he said. "I'm quitting."

One son—I couldn't tell them apart—started the machine, and threw it into reverse.

Wait, I thought. I haven't made this decision. But I am a woman experienced with men who don't take orders from women, so I bit my tongue and didn't say what I wanted to say. Instead, "Looks like good water to me. You're only down two feet."

"Not enough. Won't fill."

"But, are you deep enough to tell? Most of our dugouts are fifteen, twenty feet deep." I took one involuntary step toward the crane as it lumbered up out of the hole.

"Won't be a full-size dugout," he mumbled.

"That's okay. This one is to supplement the well over there," I said, waving over my shoulder, "so the cows don't all have to walk around that field. We figure it's worthwhile if it allows the little calves to drink; they always get crowded out around the tank."

"I've been digging around here for years," he growled. "I know."

He spat into the grass between my feet. The dragline moved up onto the flatbed behind the truck.

I was baffled; I'd been careful not to act like a bossy woman. If he kept digging, even if there wasn't enough water, the fault was mine. He'd still be paid. I was sure, from the gully's location near springs with heavy flows and from the limestone outcropping in the gully, we could get a good flow.

"Well, I imagine you've got a lot more experience than I do," I said. Maybe flattery would work. "But that one side's kind of steep; if we do get a flow, the cattle can't get down there. How about lowering that a little? Maybe you'll find water coming in from that side."

"Nope," he said, turning his back. "No water here."

"You can't leave it like this! It's a hazard to the cattle!"

He stomped away, got in the truck with his sons, and drove off. Several times in my life I've wished I had a bulldozer or backhoe, but never more than that moment.

As the roar of the truck hauling the dragline diminished, I walked down to the water, stumbling over the uneven sides of the crater. The water was two feet deep, but the hole was narrow and steep-sided. The men had created a danger instead of a viable waterhole.

I cupped my hands and drank; the water was sweet, ice-cold.

When I narrated the conversation to my father, he smiled and said, "No water. Well, we'll call it 'Linda's Folly.'" He paid the men, but he never hired them again, and he told several neighbors why.

Each time I visited that pasture, I checked the water level; the pond was full by early September. I argued that if the hole had been big enough, the water would have filled it. My father didn't argue, but he didn't say he agreed.

In the fifty years since that day, grass has grown around the pond and its edges have been softened by the hooves of the animals drinking there—which are seldom cattle. Instead, it's a secret wildlife pond, tucked away in the bottom of a gully humans rarely find. Pronghorn and deer slip through the little canyon, and the bobcat tracks are nearly lost in the crowd. Linda's Folly indeed.

August 10. Prairie grass is curing, filling the air with a crisp, light sweetness. A dozen monarch butterflies flicker out of the lilac bushes. Grasshoppers bounce out of the grass like popcorn. Two great-horned owls sit on branches on opposite sides of a juniper at the retreat house.

When I step out on the deck during the day, I can hear the drilling rig across the highway, and see it moving up and down. This driller uses a cable tool rig, the oldest form of drilling equipment, developed by the Chinese more than four thousand years ago. Still used in undeveloped countries, the machine is almost obsolete in the U.S. A derrick lifts a weight and drops it repeatedly, pounding a hole in the earth. Periodically, the driller removes the drill bit, inserts a casing to keep shattered rock from falling down the well, dumps in water, and bails out drill cuttings. Modern rigs could have done the job faster, but not cheaper. This drumming has quenched thirst for generations.

The usual afternoon thunderstorm growls, rising above the Black Hills. The cows turn as one and run, leaving calves bawling behind them. A few turn and run back, get behind their calves, and bunt them along. Lightning snaps from sky to hills and back. Highway traffic slows as rain and hail crash down. The cows slow and plod east, moving steadily as the storm pours over them.

A great reef of cloud, a white layer frothy with hail over deepening shades of gray, washes over the sky. A curtain of rain moves over the hills, and thunder makes the deck tremble. Lightning shivers through the clouds.

The sun sets, shimmering through a haze of falling hail, tiny chips of clear ice. Raindrops on the window panes turn gold. The storm moves on south, grumbling and flashing, and birds begin to call as the sky darkens.

August 11. My tomatoes and corn are thriving. Another couple of weeks without frost might bring us a respectable harvest.

We had plenty of corn in 1981, I see from my journal, as well as abundant tomatoes and cucumbers. At night, I turned on a battery radio in the garden to scare off the raccoons. George and I were riding horseback nearly every day checking cattle. In the afternoons we'd take both pickups to town for building supplies, working on the new house until nine at night, then go to bed bruised, cut, and scraped. Almost every day friends, relatives, or neighbors came to help. When we built and raised the walls, my cousin Sue and her husband Leonard pitched a tent and stayed a night or two. All that love and friendship still hovers inside the walls.

On this date in 1980 I attended my grandmother's funeral. I wrote in my journal that right after she died I felt so cheerful I shocked myself. My joy arose from her freedom from the hospital, from needles, from her memory losses, freedom to be whatever part of the world she chooses. I thought she'd rather be a chicken in her own yard than an angel with a harp. During our last coherent conversation, she said that the only important thing was to love and care for each other. A few days later, I spoke at the funeral of a young woman, the daughter of a good friend, who had been found dead in the wreckage of her car several days after she disappeared. I hope I repeated my grandmother's words.

August 12. These August days seem dusty and stifling even when the temperatures are not in the hundreds, as though the land is gasping for air.

Wanting to see my grandmother's face, I realize that I don't have a picture of her in my study and have to search to find and frame one. My mother always kept a picture of Grandmother reading in her rocking chair stuck in the edges of her bedroom mirror, but I remember her more vividly in her garden, swinging her hoe to hack a rattlesnake to bits, assuring me with a laugh that I was fine. I remember how her eyes twinkled when she said I shouldn't tell my mother that I was drinking tea: "She'd say it will stunt your growth." Hasn't yet.

Above my desk, a photo of Georgia O'Keeffe shows her striding through waist-high sagebrush, ignoring the camera; that's how I see my grandmother. My father glares from a picture taken in 1968; I know because the photographer presented it in a frame that holds a thermometer and a calendar. There are no pictures of my mother in this room.

August 13. At 5 A.M., when I wake up, I realize that it's dark. Of course sunrise has been gradually coming later for days but until this morning I hadn't been so aware that winter's hand is reaching for us. Somewhere I've read that "all matter casts its forward shadow."

Thump—thump—thump—the drill rig starts at eight in the morning even though the driller drives nearly fifty miles to get here.

As soon as we moved to the ranch, I began to notice that whenever there was a local event, certain women furnished food. Some women brought pies and cakes, others brought big roasting pans full of whatever was in season: at Thanksgiving, turkeys; during the summer, hamburger mixtures they called "sloppy Joes." After a while, my mother bought a big roaster and became one of their number until she disagreed with someone and put her roaster on a shelf in the garage.

Carrying a roaster containing sloppy Joes for fifty people into a fundraising dinner for the local history group, I realize that I have become one of the Roaster Ladies. Locally, this achievement is more appreciated than all the books I might write.

August 14. Fifteen nighthawks pirouette in front of a full moon, the teenage birds practicing their flying.

This is my adoption birthday, the day John Hasselstrom adopted me at the courthouse in Custer. I remember the kind judge who took me into his chamber and told me that if either my father or mother was ever mean to me, I should call him. My parents always made this an occasion, giving me small gifts. Now that my folks are gone, I dedicate the day to remembering how lucky I am to have had them.

Today as Jerry and I were driving along a highway in a rolling prairie, I saw far in the distance two tall spruce trees. Someone watered those trees, testifying to their determination to establish a home. No buildings remain.

What remains of our lives? Someone left those trees. With the windows open, we sang along with the radio, our voices blowing away in the dry prairie air.

Aboriginal peoples memorized details of their landscape and passed that knowledge on in songlines, a way to navigate across miles by means of story, myth, and real landmarks. Each of us creates our own personal designations.

"I'm going to the Lester pasture," my father would say, and we knew what he meant, fifty years after the Lesters sold it to us.

My journals are part of my songlines, but the names my family used for the land will vanish. When I asked a neighbor what she calls the dam we call "Lake Linda," she said, "Oh, you mean that mud hole?"

August 15. The sun rises, glowing rose-colored through a band of haze lying on the horizon. An hour later, the whole sky is the color of cold ash, the sun making the ground and grass yellow: all this because fires burning all over the west send smoke everywhere.

The well reached 36 feet today. No water yet; my house well is considered shallow at 150 feet, so we probably have a long way to go.

A full moon rose at dusk.

August 16. A gentle sunrise at 6 A.M. shows yellow evening primroses blooming before the heat rises. A claque of blackbirds rushes past, chattering and zigzagging above the grass. Goldfinches hang upside down from the sunflowers, gobbling seeds.

At midnight on this day in 1957, my mother and I came home from a 4-H dance to find my father white with pain, lying on the couch clutching his abdomen. My mother had hysterics so I drove them to the hospital—legally, because at fourteen I had a driver's license. Within an hour, my father was operated on for acute appendicitis.

My mother's journal says, "I'll never forget how I felt looking out the window at the radio towers, all the lights going out one by one as it got daylight." The lights going out: no doubt that expressed her fear of being widowed.

I remember the terror at the idea that my father might die. But I also felt happy to have been calm enough to make that drive; I knew I would survive. The next week, I moved our cattle with some neighbors helping and felt as if I were earning my rightful place on the ranch as my father's right hand. Perhaps in his right mind, he thought so too.

During this week in 1987, George and I talked with my father again about the future of the ranch. We were still getting $300 a month, not enough to live on even with George's Air Force disability. We'd tried several times to talk to my father about paying us more, about a partnership, about our buying shares in the ranch, and even about establishing a trust or conservation easement to reduce its value. He only got more and more angry. On August 18, 1987, he wrote in his journal, *"Milly called Linda. She said they are going to get jobs. We are sorry."* George's health was deteriorating rapidly, though we didn't realize how serious the decline was. He would die just over a year later. When he began applying for jobs my father wrote, *"I believe George has been in town every day this week."* I increased my speaking dates and workshops. Still, we did most of the ranch work, and we hadn't yet realized my father's mental decline.

My accountant once told me that he did taxes for a rancher and his wife who had a lot of money but filed separate tax returns. "You know," he told the rancher, "you could save $5000 by filing a joint return." The rancher refused: "It's worth $5000 not to have her know how much I make."

That attitude is a part of the reason ranchers' kids lose the ranch or learn not to care about it: that refusal to talk about the financial reality. Of course, it may also be the reason that ranchers' descendants hang onto the land even when they are barely surviving, understanding nothing of the finances but only of the connection, the bone-deep necessity to keep the place.

Because in 1987 my father would not discuss the future of this land, I now must make intelligent decisions about it without his guidance. Unfortunately, this part of the story is too familiar to ranchers' offspring in the West.

August 17. This morning I've pulled the bones from a beef stock and am letting it simmer down; packaged the granola I made last night, and done the breakfast dishes. Now it's time to write; all morning I'll see my green thumbnails—from harvesting basil—flickering over the keys and inhale that spicy odor.

I see by my father's October 1987 journal (*"Linda to check with doctor about her brain"*) that I had a CAT scan in an effort to find out why I kept having debilitating migraine headaches. The neurologist also told me I had suffered brain damage from years of untreated migraines.

When I told my father about the brain damage, he said, "Did you tell him you never *could* add?" I didn't find that funny at the time, but it was typical of the astringent humor he'd showed as I was growing up, making light of things that couldn't be changed, teaching me to "buck up."

Within a week my aunt Josephine was dead of a brain tumor; of course I had feared that I had one too.

August 18. Sunflowers six feet tall surround the shrinking pond; in the still brown water, they are reflected as a double row of golden discs. A white transparent haze covers the light blue of the sky, high clouds, too dry to bring rain. A breeze carries dust, making my nose itch.

I step out on the deck to watch a neighbor collecting cattle and am startled by *pop-pop-pop-pop.* Gunshots? When I look more closely, I see the neighbor's son on his horse at the back of the bunch. His arm rises and falls rhythmically; he's wielding a whip, snapping it correctly in the air over the backs of the cattle rather than striking them. I recall a neighbor who helped my father and me load cattle in 1977, snapping his whip on their tails every few seconds, making our quiet cattle jumpy. My father said, "I wonder how many pounds of beef that whip has cost me today."

On another occasion, another neighbor kept whipping the calves as we herded them up the chute to be branded. Our cattle usually needed only the pressure of a hand or a shout to move, but the girl kept yelling and whipping. Then she flung the whip down beside the chute and went to get a drink of water. Instantly my father snatched it up and fired it out of the corral into the tall weeds. When the girl came back, we all helped her search but none of us ever found it.

Last year on this date I froze fourteen pints of green beans and started a poem about picking beans. This supports my theory that it's good for a writer to be doing something else besides writing. Doing nothing but writing may make us guilty of too much introspection, or "navel-gazing" as a friend calls it.

HOW TO PICK GREEN BEANS

This morning's gold breeze slides
among beans slender as sunlight,
making snake patterns in the earth.
I brush leaves aside, careful
not to knock off blooms
that will make next week's beans.

Kneel
in the garden's deep soil.
Reach
to lift the bottom leaves.

Watch
for rattlesnakes.

Picking what she called a mess of beans,
my grandmother kept her hoe handy,
tilted her bifocals to see the snake,
steadied herself and chopped
until the hissing ceased.
Hooked him with her hoe, swung her arm.
The snake whirled and struck the sky.

Hold
each stem with the left hand
Pluck
each pair of beans with the right.
One hand
should always know
the other's whereabouts in rattler country.

Red-winged blackbirds sing from the cottonwoods
as I shuffle on my knees down the row.
Later, in the sinkful of water,
the beans sway like green snakes.
Grandmother used even the scabby ones,
hopper-gnawed. All winter, eating beans with bacon,
I will taste the green flesh,
taste the snake
within the harvest.

August 19. I harvest herbs—oregano, summer savory, thyme—as raucous blackbird flocks gobble sunflower seeds. At dusk last night, fifteen geese dropped out of the clouds and landed like a squadron of fighter planes on the dam. We watched them sailing around, intimidating the ducks and gobbling weeds, until it was too dark to see. After dark, we heard a honk or two as they lifted off.

The house smells of vegetation, both wild and tame, in my food dryer. When I sit at the computer, I inhale deeply and feel sure I am living in two worlds: harvesting both food and words.

Found a few ripe chokecherries on bushes near the ranch house today. I considered picking them but the "choke" part of the name is literal, referring to the extremely astringent flavor. My grandmother added apple juice to cut the acid, but even with a lot of sugar the juice is reluctant to jell. Since the grasshoppers have eaten so much of the vegetation this year, I left the berries for the birds.

August 20. The cattle spend afternoons on hilltops where the breeze deflects flies but here in the draw the hot wind sucks water from the garden. The garden plants still look butchered from the hail, but the pasture grass is still so deep that when the cows lie down to chew their cuds, I can see only sets of ears flicking: another example of how native plants can withstand the weather that is natural for this region. "I am the grass; I cover all," said Carl Sandburg.

I've been avoiding reading any journals or delving into the past for a while, both because I'm staying busy with outside work and for a break from the often-depressing thoughts they inspire.

Tonight while I soaked in a hot bath I stared at the wall covered with hand-forged tools I've found on the ranch through the years. Because Jerry does blacksmithing, I know or have been able to find out from him precisely which tools the blacksmith used to hold each piece, how it looked as it came from the forge and was hammered or twisted into shape. I can hear the hiss as it plunges into cold water.

Did the old homesteader Jonathan Sharpe make these tools? Nothing I've found indicates he was a blacksmith, but perhaps the old tools we've found were his. I can imagine the satisfaction of the maker as he created tools he needed for this ranch, knowing he would not have to buy them. He might not be able to appreciate the way I hammer words into shape for a poem, but I like to think he'd enjoy seeing that I have collected his tools from corral mud and barn dust to hang here to be seen and celebrated, another journal that cannot be read by everyone, but testimony to hard work, perseverance, and frugality.

August 21. Stifling heat seemed to rise out of the crackling grass today, as if it were ablaze, 100 degrees at noon. Still, when I ride the four-wheeler to the garden, I see flocks of blackbirds and meadowlarks spiraling up out of the grass and wheeling in great circles over the swales and gullies. They graze together on the grass seeds and insects, building up fat and strength for their flight south, away from the winter that is surely coming.

Joseph Bottum, a South Dakota writer and grandson of my father's favorite attorney, wrote that the nationwide publicity of the blizzard of 1888 that killed many schoolchildren and adults on the prairie effectively slowed the influx of settlers, who concluded that "the western prairie was unlivable." The decline of population continued for decades.

"People will stay in places where the temperature gets to 105 in the summer," Bottum wrote. "And they will settle even where the temperature gets to 40 below in the winter. But they won't stay long in places where it does both." Yet here we are.

Many rural South Dakotans have always said with pride that our extreme temperatures helps keep the population low and sturdy. A few years ago a bumper sticker insisted, "Forty below keeps the riffraff out." Money inevitably changes the equation; the influx of subdivisions of huge retirement homes into our ranching community suggests the new residents don't know the rules, that they assume electricity will always flow and that their SUVs will get them to the neatly-plowed highway. Perhaps many of them spend the winters in the south. Or else they are counting on climate change to make the winters warmer. Certainly we are due for a brutal winter like I recall from my childhood.

Recent statistics indicate that the population of South Dakota, and especially the Black Hills, is growing faster than the national average, so we may not have the protection of our extreme temperatures much longer. We need to educate people about the importance of maintaining rural areas where healthy food is grown.

August 22. At dusk, two great horned owls are calling to one another, one in the trees by the retreat house, one east down the gully.

Fires are burning in at least two places in the Hills; we see the smoke rise and then drift north or south, depending on the winds.

I picked a couple of tomatoes today, cut out the grasshopper gnawing marks, and put them in a salad.

Another realization today: I didn't get to say goodbye to either of my parents. On the occasion when I was home and my father collapsed in the dining room, he was furious when I called an ambulance and he told the sheriff I was trying to steal his ranch. When they got him to the hospital, he lied to the doctors about his previous heart trouble and hospitalizations, probably because he didn't want to die in a hospital. He got his wish: he died while I was in Cheyenne, without saying goodbye, without any discussion between us.

My mother did the same thing. She had sensibly signed a Do Not Resuscitate order at some point early in her nursing home stay. On a visit, I discovered she had been hiding the opened cans of her liquid diet in her dresser drawer; she refused to drink them. Someone wasn't watching closely. The doctor met with both of us and told her she could not survive if she didn't start eating. She seemed to understand, and I arranged for hospice care so she would not be alone. Promising to be back the next day, I went home to Cheyenne to collect materials for the retreat I was conducting the next week. I packed a black dress, just in case. She died that night.

Now I believe that she chose not to wait for me to say goodbye. The kindly hospice volunteer who was with Mother assured me that she was peaceful and ready to go—but she knew I was coming back.

I will never understand either of my parents now, and I don't need to. I can and must let them go, know they did the best they could for me and for themselves. I need to do the same for others. And my decision about the ranch needs to be based on what might be best for the ranch, not on their preferences.

August 23. On this date in 1975, I froze eighteen pints of applesauce from the old tree that stood in the ranch yard throughout my childhood. A year later, a windstorm broke many branches, and George trimmed the tree, hoping to help it recover. Instead, it died; not even the stump is left now. He felt terrible. He'd grown up learning how to prune trees in his family orchards in Michigan, but apparently the old tree could not bear the strain.

I'm sitting on the deck at dusk with binoculars, scanning the trees down the valley, searching for the two juvenile buzzards that perched on posts beside our entrance road today. They didn't fly until I'd walked within a dozen feet of them.

August 24. In 1975, my litany of harvest went on: on this day I froze ten quarts of spinach, canned a lug of peaches, six quarts of tomatoes, twelve pints of tomato soup, a couple quarts of apple-grape jelly; during the next few days I recorded ten more quarts of tomatoes, then six pints, then seven quarts. No wonder I didn't get a book published until late in the 1980s.

Some people don't realize that "living benignly on the earth" and living "in harmony with nature" takes *time.* First you have to consider the various ways a job might be done so as to do the least amount of damage. I didn't kill tomato worms with chemicals because I don't want to eat chemicals, so I used the most natural method: picking worms off one by one. If enough wild birds visit the tomatoes, or if I had chickens, the birds might do the work. Potato bugs? Same treatment, and there are zillions of them; shake a plant and they drop and hide in the mulch and dirt.

Growing and preserving food is a slow process, yet we are a culture of haste. Even when we relax, some of us do more than one thing. People apologize constantly for taking too much time, even if I say, "I'm not in a hurry." Who is ever *not* in a hurry these days?

When I look at the space where my garden used to be, I feel guilty for not using every bit of that rich soil. But I don't have the time—no, I mean I don't want to *take the time* from my writing to grow things as I used to. In those days, I knew I wanted to write a lot about my life, but I was busy living it. Now that I'm well over seventy years old, I sense the limits of my time, and I choose to spend more of it writing than gardening.

August 25. Three A.M. I'm awake and thinking. Every survey done before the highway was widened showed the traffic didn't justify the expense of widening. But asphalt never lies idle. Now trucks roar all night long, their yawl and rumble drowning out all other sound. Animals can no longer move naturally and safely from the plains to the shelter of the hills. I turn my light on to write these notes, and some passing fool honks.

Again today we visited the well site. Not much has changed; hour after hour the driller and his hired man sit on the tailgate of his truck or in the shade cast by the drill rig and listen to the bit pounding down through the layers of rock. Each time we visit he tells us about the layers through which his rig is driving: white, pink and brown clay to fifty feet; sand and gravel for five feet; more clay; more sand and gravel. He doesn't know the geological names but he recognizes the layers. He's hit water but is going deeper to be sure he has a good flow, packing gravel around the well casing and dumping in cement to help it solidify.

At three in the morning, I feel uneasy about spending the money to dig this well; my father was never willing to take on this expense. Instead, we used the pasture only when rain or snowfall filled the little dam. And I don't even have cattle; if we hit water, it will benefit someone else's animals. But what I'm doing is looking beyond my ownership to the future and improving the ranch for whoever comes after me. I consider that good business, even if it doesn't benefit me directly.

August 26. After we walk the dogs these late August mornings, I go to the tiny greenhouse to water and harvest herbs before the sun gets too intense. Breathing deeply of the growing scents is calming myself toward writing. I take the herb leaves to the dryer in the back corner of the basement. First I dump herbs that are finished drying—basil, mint, thyme, oregano—onto a clean dishtowel where I crush the leaves, then pack them into jars. Compromise. I may not grow all the food, but I can grow some of what makes it delicious. Then I spread more herbs and check the light bulbs that power the dryer. While I write in my study on the other side of the basement, the scent of drying herbs fills my nostrils.

Five buzzards sit in a circle in the neighbor's pasture. One flares his wings and rises a bit as we drive by, then settles back and bends his neck to feed on a dead calf.

August 27. Today we were headed for town, driving out our entrance road, slowly, as usual, to give the grouse and any other wildlife time to get out of the way. As we neared the botanic garden log cabin, a red-tailed hawk surged into the air ahead of us, carrying part of a dead rabbit. Jerry stopped the car. The hawk's wings flapped strongly—once, twice. He was rising toward the barbed wire fence, turning so as to fly parallel to it, but he had not yet cleared it.

The hawk had eaten most of the rabbit, but was clinging to the rest; we could see the sun through the pink flesh. The wings surged and flapped as the hawk looked back over its shoulder at us. Perhaps he decided we were too close. The rabbit dropped, and the fur caught on a barb.

The hawk turned and settled into the pasture to the north. The rabbit dangled from the fence.

Maybe he'll come back for the rest of his meal, we thought.

By the time we came home, the hawk hadn't returned. So there hangs the rabbit haunch and foot, testimony to the way nature works.

August 28. Smoke lies like fog in our valley from forest fires west of town and in Custer State Park; when I opened the windows last night it flowed into the room. News reports say slash piles are burning.

This morning, I walked through the tall windbreak trees north of the old ranch house to look at the raspberry bushes I planted in the spring—and strolled directly into a dragonfly staging area. Hundreds of them were zinging through the branches, enjoying the shade, I assume, and gobbling mosquitoes.

Computer trouble all day, so I didn't get much writing done. I try to be patient, but these glitches make me furious because I need help to solve them. When I wrote with a pen or used my IBM typewriter, I didn't need help. I know: I'm howling the perennial plaint of the elderly who can't cope with change. I don't need help writing journal entries!

August 29. Hooray! Flocks of blackbirds have arrived, beaks clattering as they march through the grass, gobbling grasshoppers. Their feathers gleam black as they swoosh away to screech in the cottonwoods.

Today in my office, I spotted another journal of my life that I can't quite read: my doll Dementina. My father bought me this doll soon after I moved to the ranch, joking that he'd never seen a dark-skinned doll before, though that's not the word he used. Even then, the term made me nervous, though all my relatives spoke that way. She's a baby doll with mahogany-brown skin, brown eyes, and ceramic brown curls, wearing a dress my grandmother made for her and holding a bouquet of real dried flowers. I named her for my best friend when I had lived in Texas, a girl who was probably of both Hispanic and African ethnicity. Of course, once we moved to South Dakota, my circle of friends didn't often include people of color.

I've wondered about the real Dementina for years. Suddenly I realize what resources I have now: and a search of the Internet turns up one person in the U.S.—in Texas—with the name. She's between 30 and 34 years old; is she the granddaughter of the woman I knew? I can't locate an address.

August 30. At midnight the wind is gusting, lightning flashes off low clouds; the kind of weather my father called "fire-starter."

We made bacon, lettuce, and tomato sandwiches for *both* lunch and dinner. A truly good BLT is only possible during a brief period in the summer when we have both tomatoes and lettuce from the garden, plus thin bacon and the perfect sourdough bread. At such times, we eat as many as possible because we won't have another opportunity for a year.

The well is finished at one hundred feet deep. For two hours, the driller pumped fifteen gallons per minute of fresh water, providing enough to water all the cattle Rick could pasture there. He'll install a pump to be left in the well and then bring over a portable generator to run it when the cattle are there. My father would be so pleased.

August 31. Evenings, two golden eagles sit on the highest point of the south ridge, looking like two people watching the sunset. Sometimes one flies for a while, then returns to the other. Finally dark hides them. I imagine them still there, watching, perhaps guarding.

In darkness on the deck, we hear the nighthawks' boom as they kill. Then something soft flutters close: a bat zings past, eating mosquitoes. Flies cluster on the screens.

When I opened my journal this morning and saw what I wrote last—"My father would be so pleased"—I immediately thought of one of his favorite sayings. I don't believe he was really a pessimist, or he wouldn't have insisted on my getting an education or continued to improve the ranch. But he couldn't help himself sometimes.

"People keep telling me to cheer up, because things could be worse," he'd say. "So I cheered up. And sure enough, they got worse."

September

September 1. Flies rumble slowly past my face, then cling to the window screen, counting out their final hours. Before sunrise, a half-dozen killdeer flew and called frantically around the pond. Were they disturbed by the resident badger? At daylight, a heron stood so still I could hardly distinguish the bird from its reflection.

We are eating well: beef raised here, with tomatoes, turnips, onions, and potatoes from our garden, flavored with herbs from the greenhouse. I feel sorry for the folks I see in grocery stores, grabbing food that is processed with chemicals, but perhaps their poor taste buds couldn't recognize the deliciousness of our food.

"A journal," writes Verlyn Klinkenborg, "always conceals vastly more than it reveals. It's a poor substitute for memory, and memory is what I would like to nourish." The journals I've been reading this summer have recalled many memories I might have lost without them, and yet I've been conscious of what those journals were not telling me.

So here's another question: what am I concealing as I write this journal? Am I still hiding important truths from myself?

September 2. The 3 A.M. coyotes are right on schedule, howling and yipping just as the morning's first breeze eases in the windows.

Once I'm settled at my desk facing the basement windows, I see the constant flicker of young and old barn swallows zipping under the deck, flashing in front of the windows as they catch flies.

Upstairs, checking on lunch, I see a billow of gray smoke rising, grab my phone and start calling to find out where the fire is. A neighbor's hay, stacked too green, has spontaneously combusted. Neighbors come from all directions to fight the fire but by the end of the day he has lost all the hay he worked hard to harvest for winter feed.

This has been a busy week; I read and commented on a 140-page manuscript, planned three retreats, made six pots of tomato sauce to put in the freezer for winter, and read six mystery books as well as doing my usual work of making three meals a day, watering the garden, writing letters and emails. Seems the world keeps spinning faster.

When I feel overwhelmed, I may walk out to one of the gardens or on the hillside with the dogs, deliberately looking for the materials for a tiny bouquet. I select a few small blooms, thinking of nothing but their color, texture, size. I put these in one of several small vases that I place directly above the kitchen sink where I will see it often during the day.

With the bouquet, I create a tiny island of calm in the middle of hurry. Every time I look at it, I breathe as deeply as I did while choosing it. For a few days, each provides considerable balm. Sometimes I add the dried flowers to another arrangement.

In the same way, when I'm too busy to write—which seems to happen much more often than it should—I may take time to deliberately create a paragraph with no particular purpose. Most often I do this when I wake around 4 A.M. I switch on my reading light and pick up my journal from the bedside table. If I can keep the dogs from leaping up and running downstairs for their first morning outing, I can write quietly for a half hour, especially if the highway is empty so I can hear wind through the grass, a wind chime on the deck.

What I write may become part of a longer piece, or it may be just a morning reflection that remains in my journal. Either way, it helps me begin the day in peace.

September 3. At 7 A.M., the sun is two fingers above the horizon, gilding everything in pale gold with green highlights. A small plane chugs past our roof, headed north.

Nature tells us in dozens of ways the summer is ending: black and white dragonflies arrive to eat the mosquitoes. Pollen turns the air yellow and makes the breeze an enemy.

Our senses warn us about the end of the hot season; wasps zig, zag, and *zizzz* around the screen door, occasionally slipping inside to bump against the ceiling. Ignore them and they will surely land on my shirt collar. Coyote pups howl shrilly just before dawn. Birds lift in dark flocks from one field to alight in another, gobbling seeds. At a nearby garden center, we smell roasting hot green chiles, and buy a bushel to peel and freeze. Meadowlarks stomp through the drying grass, tilting their heads to peck up seeds. Two red-tailed hawks glide over the pond as the ducks tip their tail feathers in the air, gleaning something edible in the muck.

As a canny gardener, I nip off the new shoots and blossoms that don't have time to become tomatoes, encouraging the plants to put their energy into ripening larger tomatoes.

In the evening the bats chitter in the bat house over the garage door. Cranes have been going south for several days, whooping and crooning as they circle higher right over the house. If they catch a thermal while we watch, they shoot south so fast they leave us breathless. After dark, the coyotes sing counterpoint with two great horned owls.

September 4. I've read in news stories that if people stay home instead of going on vacation, it's called a "Stay-Cation." The west porch of the retreat house and the flower beds beside it have become the "Stay-Cation" spot for a flock of grouse that have been growing up around the ranch yard all summer.

Last year I purchased and planted ground covers along the house foundation to anchor the dirt. They died over the winter, so this spring I transplanted native ground covers—goblin gaillardia, several kinds of stonecrops and some violas—into that location. They thrived all spring and summer. As fall approached, I spent one evening digging up more goblin gaillardia and oxalis from the gardens around my house and transplanted them to the bare spots.

The next morning when I went to water them, the plants were gone or lying in the dirt, shriveled, among bowl-shaped depressions in the loose dirt just like the dust baths my chickens used to make. Brown and cream feathers decorated the spot. Grouse!

I transplanted more gaillardia, more oxalis and stonecrop.

The next morning, not only were the new plants gone, but the west porch and step of the retreat house were liberally decorated with grouse droppings. I can picture the teenage grouse gathered there in the cool evening, dusting themselves, perching on the outdoor couch, and sunbathing on the concrete steps as they chuckle and gossip or whatever young grouse do as they prepare for winter.

I retrieved a few surviving stonecrop plants and tucked them inside the window well, hoping they will survive until spring. Pondering my choices—I'd rather keep planting than to harm the grouse—I notice the birdbath my father made for my mother. He built it where she could see it from the two picture windows in the house. A shallow basin made of concrete poured into an iron wagon wheel stands on an eight-inch-thick column of concrete reinforced with old iron. Mother filled it daily with a hose and loved to look out at it. In winter, robins, grouse, sparrows, and other birds still live in the yard, and cedar waxwings visit every fall to gobble juniper berries. My father was not a deft builder, but he created the unwieldy thing with love.

At dusk, I'm sitting on the deck with a book, keeping watch on a thunderstorm hanging overhead. The air is still. The cloud above rumbles low, like a dog growling in its throat. The air around my face stirs slightly, as though I am lying next to someone sleeping. The cloud rolls and heaves, black and wicked looking.

Then a single nighthawk flies across the valley from west to east, its thin *peent peent peent* bouncing off the clouds.

Cosmo noses a crack between the boards of the deck, and I look down to see wasps surging up out of a nest we didn't know was there. I go below the nest and squirt it several times with my dish soap and water mixture, dodging wasps as they die and fall. I realize that we attract wasps by building structures where they can anchor their nests. And I know some type of wasps do us good, killing insects we like even less, like caterpillars. Some sources say that without the carrion-feeding habits of wasps, we'd be knee-deep in dead insects by summer's end.

September 5. This morning I notice Jerry peeling potatoes for breakfast hash browns. He uses a potato peeler and slices forward with short strokes. I peel potatoes by slipping the peeler under the skin at one end and drawing it down the whole length of the potato.

How did he learn? He can't remember. At first I think I learned from my grandmother, but then remember she used a paring knife, starting at one end and peeling an impossibly thin strip of skin down the length of the potato. Once I discovered a potato peeler, I never went back to the paring knife because I could never get the peel as thin as hers. She could peel an entire apple in one long strip with that paring knife, its blade worn thin by years, leaving the skin coiled beside her plate. I picture a half-dozen women, each reflected in the eye glasses of the last, stretching back into generations, all peeling apples the same way. How far back, I wonder, could these physical habits reach? Here is another kind of journal: the knowledge that is passed down to us from ancestors and others, and which becomes part of our lives even if we can't remember its origin. If Jerry had a son, would he imitate that way of peeling potatoes?

As the sun sets, the prairie colors become even more vivid: each grass blade glows green or gold; eleven antelope are outlined against the grass, their bellies so white they look like crystal. The long red building near the barn shines as vivid as if it were lit from within; the grass in the yard is electric green.

My mother's journals have been piled on a table in my office for weeks. Every now and then I open one and try to read a little. She wore her wig to church but writes, *"I didn't fool anybody."* She narrates the plot of a TV show she watched. On a scrap of paper I try to write as small as her handwriting in these journals, as small as the type on a printed page. My hand cramps. I hadn't considered how writing that way might have affected her hands. Like my grandmother, she used her hands hard: peeling potatoes, washing clothes by hand in cold water. While my hands, like my grandmother's, show swollen knuckles and are beginning to be shaped by arthritis, I don't recall my mother having trouble with hers. She had spent years as a secretary, taking shorthand notes, typing on the old typewriters that required so much strength to use. These days, those who operate computers, which are infinitely easier to use, have carpal tunnel syndrome and related injuries. Mother didn't. Or maybe she simply endured the pain, since she had no name for it.

Should I burn my mother's journals? I can't stand to look at them anymore, but if I burned them, no one else could ever examine them or learn from them. I believe we need to learn what we can from the past. Might some future make my mother's journals useful to someone?

I can't take the irrevocable step of destroying Mother's journals, but I yet don't know how anyone else could read them. I thought about it while making both peach cobbler and cheesecake.

September 6. At sunrise it's 55 degrees and raining. Making tomato sauce and snapping the ends off the green beans makes me think of my grandmother. As we sat on the big rock that made the step up into the porch of her little house, she told me how to snap each bean in two or three pieces. At first my fingers tangled in the strings. Today's beans seem to be stringless; another of those changes of which we aren't really aware.

Today I dug into the shoebox of my grandmother Cora's letters to my mother. There's only one from 1930, written to my mother on August 23. She reported that Walt was still doing some car repairing, but he had closed his shop, probably because it wasn't paying expenses. Cora said he would *"probably have to put an ad in the paper next weak and open the shop a gain that will not be very paying business especially for winter but will help some we shurley would like to sell the place here we would get out of debt and go somewhere Else for a while or for a change. . . . I don't think I have ben so Blue and discouraged for years and years."*

I can hardly imagine my grandmother discouraged; she must have learned so well over the years to hide her sorrow that I never saw it. She always seemed not only cheerful but calm in the face of any crisis, able to expect the best, and make do when times were difficult.

The dates of these letters are scattered; no doubt Mother threw some away. In 1940 Grandmother reported that she'd taken a driving lesson from Dad so she could get around a little more by herself, but she was never very comfortable driving. In December of 1943, she wrote of a neighbor who was *"taken to Yankton,"* the state mental health facility. She was *"going out in the nite and plundering around for hours in the cold. . . . She taken the car and drove all around it's a wonder she didn't upset the places she went."*

On February 10, 1944, Cora wrote, *"Every time I go to town for something to make Linda dresses and pinafores—maybe I can have something for you to sew if you get home this summer if I don't have time and it doesn't look as tho I will get to do mutch a long that line."* I'd been born in July of 1943, so I was barely seven months old.

Grandmother had started a letter, saying, *"This is your weakly letter My Dear"* when her life changed forever the day her husband, Walt, died. She wrote, *"then this tragic thing happened to us at 12:30 while we was up to do our feeding it just so happened that Goral* [a neighbor] *was going by whare we was going to load the truck or hay rack and he brought his back and came over and helped us load and was just visiting a few minutes in 10 minutes more he would have gown on and we would have been scattering our load he always drove and I would be up on the Rack scattering the hay just all at once he half turned to me and said, 'My God, I am going' then droped and in 5 minutes time he was gown it just looks like it was to be that way he had started to build me some cupboards had a sink and everything bought to fix my kitchen corner he turned the windo over the long way so he could put the sink under it and was in joying doing it at odd moments but he never got it finished."*

George and Harry were in the military, so she writes, *"we had the Red Cross send the boys messages they were sent Sunday and we had an answer from George this eve and we are going to try to get George home to take over since he had served so much time over there I believe we can."*

Later that year she writes that the family that could get home for the funeral had enjoyed the box of chocolates my mother got every year for Walt's birthday. *"We all know,"* she added, *"how our dear Dada had come to expect a pound of the best chocolates,"* but that *"he had gotten so the past 2 years he ate very sparingly of it to the side of years before."*

On April 21, she wrote, *"My life seems so empty and the future looks so dark. . . . Then something seems to brace me up and give me courage to go on just a while longer."* That's the grandmother I knew, always able to find something to make her go on. She taught that skill, that grace, to me, somehow. One of her sons, Bud, was home and though he had a family in town, he came to the ranch every day to help with the work. She was milking three cows and selling the cream for 50 cents a pound. Her other two sons were still in the service and though she wrote to her congressional representative, George was not discharged for more than a year after his father's death; Grandmother braced herself up and went on.

The shoebox holds only a few letters from the 1950s; surely Grandmother would have responded to Mother's news about her dating experiences once we moved to Rapid City, but Mother must have destroyed those letters. By 1964, I was in college and once a week she wrote to me as well as to my mother.

In December, 1964, she wrote to me, *"My mother always said the les one has to do the les they want to do and I have come to realize how true that is in fact many of her suggestions come bak to me now days. Folks in her day was a lot more intelligent than I used to think they was."*

Yes, Grandmother, and you would not be surprised to know that I now realize you were a lot smarter than I thought at the time too.

Once I took my 1954 Chevy to college in the eastern part of the state, she suggested that rather than drive home on slick roads I should take the bus; it would cost me about $15 she said, and I couldn't drive it for less. She was right, but I'll bet I drove home.

Only a few more letters remain in the shoebox. On July 1, 1966, apparently responding to my complaints about the cost of new furniture, she recommended second hand, but added how *"awful it is to clean up someone else's dirt."* Thanks in part to her suggestions, I became devoted to secondhand stores. At the end of the letter she said, *"The years go by so fast live today my dear ones."*

In January 1977, she wrote, *"I do very well always something I can do if I happen to look at my self in the looking glass I look like an old laidy in her 80s guess I have to get rid of all looking glasses anyway I am not sick and can get a round very well for 84 years thankful I can do as well as I can could be worse."*

When I asked my uncle George if he'd like to read her letters, he said, "She never said anything anyway. Oh, maybe what she was doing but nothing about what was going on in the world."

I disagree with my beloved uncle. History books can tell me what was going on in the world. Her letters reminded me most of what we do in the world is ordinary and provided evidence of her steady love, declared every week no matter what her letter said, and the wisdom she gave me with her advice. Priceless.

During that year, she began repeating herself in her letters and telling us each time she wrote that she was *"Cora B. Hey."* At the end of one, she wrote, *"Just an ordinary mother."* She died in 1980 at the age of 88. Her picture hangs over my desk and her smile greets me every morning.

September 7. *"Our monarch is in a chrysalis,"* reads the last line of my journal for Labor Day 1981. As we were putting up the roof rafters of the house we noticed a monarch butterfly perched at the very peak and worked carefully around it for two days, until we realized it was encased in a chrysalis. When it was time to put plywood over the roof, George carefully dug out the sliver of wood the chrysalis was attached to and put it somewhere—but where? Perhaps in one of the windbreak trees, tiny then, but large enough to shelter a butterfly. We didn't notice whether it hatched safely, because we were too busy building our own chrysalis.

On the same day seven years later, George died—shed his diseased chrysalis and spread his wings.

Will the monarch survive in the world? In recent years, we've seen a few monarchs here; they love dill so I plant a row or two every year. My favorite botanist tells me that several species of milkweeds are native here, including *Asclepias verticillata;* these are short and with skinny pods, not the tall, showy milkweeds of areas with more moisture. So perhaps the monarchs we see here benefit from this native plant, which is abundant on this hillside. Worldwide, monarchs are declining. In 1996, the year Monsanto introduced its first Roundup Ready crop, monarch butterflies covered 45 acres of forest in Mexico, the equivalent of 34 football fields. In 2013, the butterflies covered just 1.65 acres. The chemicals were not used on our pastures, yet the tall plants with their showy pink flower clusters have vanished, probably because the county sprays weed killer along our roads.

September 8. Students from an Iowa college are visiting the ranch this week to learn about the prairie and writing. Today, I walked with three of them to the railroad tracks to look at the old cars the railroad companies used in the 1930s to stabilize the embankment. In the ditch was a stagnant pool and we spotted four garter snakes zipping around in it, catching tiny minnows. Each snake slid into the water, moving sinuously across the bottom. We watched every curve and thrust since the water was barely deeper than the thickness of the snakes' bodies, about three-quarters of an inch. When a snake captured a minnow, it raised its head from the water and gulped.

Later that day, several students and a professor waded in Battle Creek. Sitting on the bank, I mentioned that when I was nine years old and played hooky with my cousin Roy, my parents found out because as I took my bath that night, I found leeches between my toes. I called my mother, who threw a towel over me and screamed for my father. He squashed the leeches while quietly mentioning that he got them when he played hooky too. Horrified to see my blood squirt, I was more distressed to disappoint my father, who made it clear I must not play hooky again.

When I mentioned leeches, everyone leapt out of the creek and started looking between their toes: yes! One girl found a leech, not yet firmly attached.

I was glad to know that while some things change, the leeches still inhabit Battle Creek. I wonder if students still play hooky and discover them as I did. And if the monarch butterfly doesn't survive but leeches do, what does that mean about our world?

My father's journal for 1988 reads, *"Low 46 degrees, high 80 degrees. The funeral for George is this eve at 6:30. I disked some creeping jenny. Transplanted one cottonwood tree."*

What a reminder of the way emotion was treated in our home. My mother wept and wailed that I was a widow and would always be alone; my father worked, his antidote for everything. My uncle George, my mother's half-brother, came with his wife, Nancy, to the ranch to ride to the funeral with my parents. My mother, of course, wasn't dressed. He fixed a door latch while he waited; another practical man.

In his journal, my father mentioned how nice the outdoor graveside services was and added, *"I felt much compassion for George's mother."* He didn't mention me. But he planted a tree.

September 9. Today I woke up thinking of my poem about learning to breathe and of the damage done by refusing to acknowledge emotion. Surely my father taught me to do so by his example. People have so often written about the "taciturn" Western hero. Heroic, maybe, but definitely unhealthy. Yet many people subscribe to the idea that hiding emotion is a good thing. This poem has taken me decades to complete, though at the same time I've been writing other things. My mother's journals have helped me finish some thoughts. Perhaps by putting the damage into a new creation, I've both made it more powerful and robbed it of its power to hurt me.

BROKEN GLASS

She found more whiskey.
That's how it started every time.
When he came home
she screamed and
he yelled. I was three,
crouched under the table
holding my breath
as she broke bottles

in the kitchen sink.
I could see his ankles,
shoes set wide apart facing
her hose and high heels.
Smash. One. Scream. Two.
Sour whiskey fumes choked me.
Glass shards pierced air,
shrieked against the tile floor.
Three. Pop. Four. Bash.
Holding my breath, I counted.
His drinking, her spending.
How he left me alone while he bedded
the woman upstairs and now
she's having a baby. If I
held my breath, they'd stop.

That night mother carried me
up steps that clanged
onto a chugging train.
I held my breath and counted
lighted cars uncoiling
behind us in the dark.
Mother divorced father,
found a job, married a good man.
When she slapped me,
I held my breath and counted.
Her good man died. She
shriveled away into eternity.

For sixty-five years I've
held my breath and counted.
This poem is me learning to breathe.

I recall visiting with an editor at a major publishing house about the manuscript that eventually became my book *No Place Like Home,* brought out by another publisher. After I'd mentioned that I included some essays about George, she leaned forward, patted me on the knee and said, "We've heard enough about your dead husband, dear." I've since spoken with many authors who consider her a brilliant and compassionate editor, but her comment made me think I shouldn't write about George. Yet learning to live and laugh in spite of death is surely one of the most important lessons of maturity. Surviving George's death has made my own life more precious and made me appreciate my good fortune in having a loving partner. Many widowed readers have told me my story helped them.

Perhaps one of the reasons some people are unable to consider death as a necessary part of life is because "we've heard enough" about it; we're supposed to "get over it." Of course I regret George's death. I relive every moment of his last illness whenever I re-read what I wrote about it in my book *Land Circle.* But how could I wish him alive after he became paralyzed? If my life had stopped with his, what a lot of joy I'd have missed, including the chance to come back to this house we built together and live in it as we believed we both would do. Jerry and I enjoy each day filled with work we have chosen, aware of the lives flowing around us, and aware of those who aren't here to enjoy life.

I honor George by living.

Then comes one of those moments when I realize that my emotions are not unique. Just before Christmas in 1945, when her two sons were able to come home for Christmas, my grandmother Cora wrote to my mother, *"I dreamed of Dad last night I thought he had ben to Rapid an somewhere and drove up in the Pickup and he was dressed up and when he came in the house I had broken my glasses rim and he taken them and was fixing them for me I still always miss him so how he would injoy this Christmas with the two boys at home."* She dreamed of her dead husband helping her, as he had in life, and as I often do.

Lovely slow rain this evening in three separate storms; filled our rain tank. Now I'm sitting on the deck in the last light watching the cows and antelope graze, ignoring each other. I read a little for recreation after reading the writing of retreat participants all day.

September 10. At sunrise, fog lies low. The vivid yellows, greens, tawny browns of the grass were washed clean of dust by a quarter inch of rain at dusk: silver rain falling from a silver sky. Now the sullen sun crouches behind waves of gray.

I tucked my grandmother's letters neatly back in their shoebox today, labeled it, taped it shut, and took it to the shelves in the basement of the retreat house where I am storing other archives until I decide what to do with them.

Later, I got out my old portable Singer sewing machine and was rendered speechless by discovering a tattered and faded creature tucked inside. I started sewing when I was about nine years old, after we moved to the ranch and I joined 4-H. My parents bought me a cabinet sewing machine and installed it in the guest bedroom where I sewed my way through high school. I remember my terror in the "Make It Yourself With Wool" competition, which required contestants not only to make the outfit but to model it for a crowd consisting mainly of the parents of other sewing models. I can still feel myself walking stiffly up the aisle of the Congregational Church under the eyes of half the town, including the sewing judges who watched to see how well our clothes fit. The winner was a high school boy who made a stylish Western jacket; no one derided him for sewing.

When did I get my portable Featherweight Singer? Probably when I left home for college; it was old then. I see similar models for sale online for several hundred dollars, but I won't sell mine as long as I can thread the needle. I've sewn everything. I made elaborate pajamas and nightcaps for my first husband's daughters when we were too poor to buy gifts; I've made suits and coats, stitching my way through grad school and both marriages.

Wherever my little sewing machine went in its sturdy case, it carried this odd creature along. I had originally found it in grandmother Cora's sewing basket after her death. I'd forgotten I had it until today. An ancient piece of cardboard stiffens the silhouette of what appears to be a penguin viewed from the side. The cover is blue fabric with pink roses, while the inside is a simple blue stripe on white. The edges were folded neatly around the cardboard and whipped together with tiny stitches. A black cross was stitched rather sloppily on the outside in the position of an eye and a white button eye was sewn inside. A pin cushion.

Did my grandmother make it? My mother? It has to be more than a hundred years old; the fabric is faded and so thin I keep only a few pins inside.

So what is the Pin Penguin's worth? Memories. I can see my grandmother's old hands, as age-spotted as mine are now, holding it as she replaced a needle I'd threaded for her. Another page in her unwritten practical journal, like so many of the diaries of the women. Now she, and the penguin, are part of mine.

September 11. A ruffly, flirty hussy of wind strode among the grasses today, her fingernails clicking like razor blades, reminding me of the ticking of ice chips against the windows. All day I hurried to harvest more corn, to put tarps by the tomatoes so I could cover them if frost is predicted. Around me, I sensed the ghosts of early settlers and Indians, all storing up food, recognizing that sharp sound as winter's approach.

My grandmother and her pin penguin were waiting for me when I woke up. Stitching on my faithful old Singer, I thought no more of writing than she did when she was sewing. With the hum of the machine came the calm of accomplishing something good, a practical gift for a man who is so good to me.

I made Jerry a cover for his flintlock rifle. which is so long no modern gun case will cover it. We've been wrapping it in a sheet, but I realized today that three legs from old jeans would make good protection. The little Singer sewed right through layers of denim without a hiccup.

My grandmother must have felt the same way as she finished sewing the doll dresses she made for me or mending her husband's jeans or making a dress for her daughter. I relaxed into the job and let the email, flashing with self-importance from my computer, wait. I began to think about the history of the pin penguin, about my grandmother, the work she did.

I realized that this is my favorite way to have writing "happen." The sewing produced the necessary solitude and calm that produces writing. This is why writers must not fill every day with schedules and activities. We need to leave room for the unexpected, the thought that begins in some task, perhaps, and blossoms into something more.

I'll cherish my grandmother's pin penguin even more for this new gift.

September 12. An hour after the sun rises, a rabbit is dozing in a pool of warm light against the propane tank.

Our resident coyote is a big one. This morning when I spotted the tawny shape moving across the hayfield south of the house, I had to grab the binoculars to be sure I wasn't seeing a deer. The coyote trotted to the edge of the pond and eyed the gathered ducks, who all raised their heads and became very alert, but remained silent. The coyote watched for a while, then turned and trotted up the hill, heading toward the badger dens on the ridge.

Every day I receive email or mail notes from fellow writers, telling me how many classes they are teaching, where they are going to read next, what they are doing to help the local school system improve students' ability to read and write. Sometimes as I read a mystery in the evenings, or take time to bake bread during the morning, I feel guilty that I am not working harder for writing or education.

In part, these activities and jobs arise from the fact that a writer usually has to teach or give "presentations" or in some way "promote" himself or herself in order to make money. Most of us are not supported by the sale of our books, even if our names are well-known. I did a lot of these things. I taught endless classes to unresponsive students for little money. I talked without charge to dozens of groups who styled themselves as helping the public and promised that, instead of being paid, I could sell my books after my talk. Perhaps some of that hard work did some good, but it was not writing; it took me away from the writing I needed and wanted to do.

Moreover, most of us have been made to feel, as I was, that writing is not "a job," that it is somehow not respectable. "Making any money at your writing?" my relatives always ask. The implication is clear: if you aren't making money, you must not be a successful writer.

Writers write. Unless they spend time alone, where writing may occur, they are probably not writing to their own satisfaction. Taking time to sew, to handle Cora's penguin, reminded me once more that writing requires time spent away from the business side of writing, time spent simply thinking.

Publishers and some writers tell me I need to be giving more speeches, "promoting" my work more, but I believe that my time is best spent writing as well as I can. I hope that publishers will promote my books so that readers find them. This may not work, but I grow less and less able to do what is called "promotion" no matter what the cost.

September 13. Reading in my chair beside the window so I could hear the rain, I saw a movement: a wet kestrel was shaking himself on the deck railing. When I turned my head, he saw me through the window, did a double take worthy of a slapstick comedian, and shot away.

This seems almost too silly to record, but I am finally breaking free of the behavior I learned from my parents, to keep everything because "it might come in handy some day." I've spent years discarding, giving away, recycling, and in some instances burning the things they kept as they followed that philosophy. Yes, some of the items did "come in handy," but a huge number of them were just a burden. So I've cast a critical eye on my own possessions. Today I made a pile for the second-hand store: four lovely wool jackets that have been too large for me since the day I bought them; two pairs of shoes, also too large, so they were causing me to trip; and several pairs of pants that are also too large since I've been losing weight. If I gain weight, I'll buy more pants.

I'd kept my mother's wheelchair, telling myself that I might need it, but when I sat down in it, I realized it was too small for me. I took it to a nursing home in Custer, where the delighted director said it would fit a woman who couldn't afford one. Cynically, I remind myself that my altruism wasn't entirely unselfish; if I need one, better ones will be available.

September 14. The barn swallows have vanished, gone south I suppose. Easing into the retreat mindset after breakfast, I picked a little fresh arugula and a tiny yellow squash for lunch. As I gathered hollyhock seed in a huge chicken feed bag, I noticed the shed skin of the big resident bull snake among the rough stems. Bees are still mining the hollyhock and evening primrose blossoms, buzzing deep inside the blooms and helicoptering out covered with pollen. The squash plants are shoulder high, yellow blooms looking like square dancers' skirts. Every now and then I find a huge and aged yellow squash hidden deep in the foliage and put it aside for my neighbor's chickens.

Beside the east deck, the chives and garlic chives are sprawling and blooming, and the violas I transplanted a few weeks ago survive. Some of the anise that is taking over this herb bed stands nearly as tall as I am, its purple candle of bloom scenting the air. The columbine has bloomed and tossed its seeds onto the ground, so I leave the hose nearby for a few moments. When I go to the retreat house, I find a beautiful brown thrasher dead on the step. Perhaps it flew into one of the windows, though we keep blinds over them partly to prevent such mishaps. I place the bird on a high branch for burial in the elements. Then the current retreat writer and I sit at the round metal table beside the lilac bush and return to our conversation about writing in the natural world, our talk reinforced by the dialogues of birds and leaves over our heads.

One more memory of my grandmother surfaced this morning, because I gave myself time to think while gardening: some of her expressions. In situations where my mother swore vehemently, my grandmother usually said, "My stars!" When I'm tempted to swear these days, I try to remember to say instead, "Heavens to Betsy!"

September 15. This morning I woke from a dream in which I had been in the basement of the Congregational Church in Hermosa, which I attended as I was growing up. The place was crowded. I was looking for my mother, walking through the basement that was much larger than it really is. In one room, children were barefoot and a couple of women were pouring water on the floor and explaining that the children were to be baptized. The women had already prepared bowls of richly-scented spices; I could smell cinnamon, nutmeg, cloves. I saw my mother walk by in the hall so I left before the ceremony began, but she was out of sight. I hurried down the hallway, looking over the heads of people, pushing through the crowds as politely as I could. Every now and then I would glimpse her walking somewhere ahead of me, but I never caught up.

I never caught up. Never understood where she was headed in her life. Dreams sometimes tell us what we are too blind to see in reality.

September 16. This morning we were watching the heron as we often do while having coffee in the bedroom. The bird raised its head, holding in its beak something very large and shook it violently. We looked through binoculars. A turtle? No, too limp. A plant? Too large.

The heron raised its beak as high as possible, lifting the object free of the water, then shook it hard a few times and dropped it, then picked it up again, and slammed it to the ground. Finally we concluded that prey was a dead duck, and perhaps the heron was ripping off rotting pieces and eating them. One source we checked reported a heron killing and eating ducklings. Experts say they may use their sharp beaks to impale small prey before gulping it down.

September 17. A little fog at sunrise, a few drops of rain, reminding us that this is autumn with winter coming. Pulled some tomato vines and hung them in the cellar; I've heard the tomatoes will ripen that way, but I'm skeptical.

Today I was sorting photographs again, wondering why I think of my grandmother when I first look at myself in the mirror in the mornings. Suddenly I noticed how long and high my mother's nose was. In some of the photos of her as a young woman, it might almost be called Roman, a nose with a high, long arch, with the kind of character I've wished my nose had. With her long, dark hair, she looked like an Italian princess or a movie star.

The picture of my grandmother over my desk shows *her* nose was long and straight, her nostrils growing broader and fleshier in her eighties. Now I remember how both of them teased me about my "pug nose" when I was a child. At my mirror I realized that my nose is much shorter and my nostrils are smaller than either of theirs.

Hmm. More pictures. My biological father's nose was huge, flaring out from his face like a Gothic arch. My uncles, children of my grandmother's second husband, have noses I think of as English: chiseled, small, sharp, as do their children. Finally I look at a photo of my grandfather, Harry Elmer Baker, Cora's first husband. His nose is exactly like mine: small, short, nondescript, but definitely the same. Suddenly this grandfather I never met becomes a real person, blood of my blood, telling his story in yet another kind of journal: flesh.

In the photo above my desk, he stands solidly in front of a horse named Pedro. His left arm is cocked, hand on his hip; he wears a white shirt and open leather vest. But the reason for the photograph, my grandmother said, was the magnificent pair of white sheepskin chaps that cover his pants and flow down over his boots: "He was so proud of those chaps." They were married in 1907 when she was sixteen and he was nineteen. My mother was born in 1909 and her brother in 1912.

He was nineteen in the photograph with his chaps. Beside his picture is one of my grandmother, seventy years after their marriage. I stand on my desk for a closer look. Elmer's broad-brimmed hat tilts back, showing an unsmiling face in which I see pride and perhaps even arrogance. And the nose that I see in my mirror.

Is this why most people are so anxious to reproduce? Perhaps the desire to see one's own face staring back from a child is instinctive, part of our biological imperative to continue the race. That particular strain of DNA must have slipped out of the mix in my case; I've never wanted to see myself looking back at me from someone else.

September 18. Two hawks circle over the house. I grab my binoculars to look more closely; yes, they are harriers, slim and gray, with long tails.

When they swoop low, I can see the white patch at the base of each tail, how the wings make a V.

An hour ago, I saw one of them dive at the other, screaming, and the first hawk dropped some small meal, probably a vole. They both looked as if they were diving for it but I lost sight of them behind the windbreak fence.

Walking under the pine trees my dad planted on the north side of the ranch yard, I am puzzled for a moment by the sound of the wind through them. The sound is completely unlike the wind through the junipers or the cottonwoods; it catapults me to the back yard of the big log house my grandmother's second husband, Walt Hey, built in Elbow Canyon. Hanging up clothes there with her, on the old wavy clothesline, we were surrounded by that *whoosh,* those pine needles fingering the strings of the air. Somehow, even on this plain, the wind through the pines is the sound of mountains, the sound of distance. That reminds me that my uncle George has lived in that canyon ninety years, listening to that sound. Well, of course he went away for schooling and military service, but he returned as soon as he could to that little ranch, took care of my grandmother, married, and reared his children there.

Now he cannot hear that sound of the pines because a forest fire burned the trees around his ranch. My uncle insists forest officials let the fire get out of control because they wouldn't let neighboring ranchers fight it. Instead, they took time to study maps before deploying firefighters unfamiliar with the trails and pasture fences. Eventually, they evacuated my uncle and aunt and barely saved the house. The canyon slopes are covered with blackened pine trees, and the wind no longer makes that murmur he's been hearing, awake and asleep, for his entire life. He has always been cheerful in adversity; now he is gloomy and says he is waiting to die.

September 19. Drizzle at 5 A.M.; I made green chile and apple crisp for lunch.

In 1982, my father was 73 years old; he'd begun pasting or copying poems into his journal. Today I found a few lines he'd written from memory and called "The Man in the Glass." The poem ends:

> You can fool the whole world down the pathway of years,
> And get pats on the back as you pass,
> But the final reward will be heartache and tears,
> If you've cheated the man in the glass.

After all the searching I've done for details of my father's life, perhaps this is all I needed to know about him: that in his twenties he memorized that advice. On the wall above my computer, he stares straight out of a photograph, his old plaid shirt unbuttoned, his damp hair streaked by the comb, and looks without flinching at the man in the mirror.

When I was first thinking of myself as creative, I was surprised at the poems he'd memorized. He'd recite them as we drove the long miles over east in the pickup because he wanted me to find the complete poems for him. I recall his rendition of:

Strike the tent! The sun has risen.
Not a vapor streaks the dawn
And the frosty prairie brightens
To the westward far and wan

This stirring, romantic, overblown poetry was in style during his school years. Now that I am nearly his age, I entertain myself by reciting the much fewer poems I know, and regularly deliver a diatribe about what a shame it is that students are no longer required to memorize poetry.

In my father's journal I find the same lines about the "moving finger" that my mother copied:

Nor all your piety nor Wit
Shall lure it back to cancel half a Line,
Nor all your tears wash out a Word of it.

They enjoyed reciting those lines, especially when I was in high school and had stayed out too late the night before or had embarked on some project they disapproved. Even if I destroy their journals and my own, I know, as they did, that nothing I can do will cancel what the "moving finger" has written.

September 20. The fall equinox is traditionally a time of thanksgiving for the harvest as well as of sorrow for the coming darkness; day and night are balanced on the tip-top of autumn before we fall into winter. Finishing the old business of summer becomes an essential part of preparing for winter, when we will hunker down into rest and relaxation from the hot season's demands.

In the garden, Jerry helps me pull the markers and water pipes from the rows where we grew radishes, turnips, and squash; we'll store the water system in the garden house for next summer.

In my garden journal, I make notes about what went well and not so well during the past months. As a child, I was taught to grow everything we possibly could in the garden; at my age, I'm still trying to wean myself away from that idea. Now I admit to growing more tomatoes than we need; I'll reduce the number of plants next year. Turnips: we don't especially like the type that grows best in the garden; we'll buy what we need. I'll plant zucchini, so we can make *calabacitas con queso* and because it's easy to give the excess to friends. Pumpkins: no. Jerry's diabetes diet rules out pie and even the pumpkin bread I love to make. I'll definitely grow the herbs I love: basil, oregano, parsley. I may experiment with more herbs since, with fewer tomato plants, I'll have more room in my raised beds.

As we close down the garden, we also stock up, harvesting. This contradiction seems essential to autumn: pare down, stock up. Hot days, cool nights; bask, shiver. I enjoy the feel of the sun on my back for the first time since spring, but relax in a cool breeze on the deck now that the mosquitoes have vanished.

September 21. The most recent retreat guest left this morning. We did good work, but it was a difficult few days because she had a severe cold and was depressed. So as I collected the sheets and towels, I started a smudge fire of sweetgrass and sage and carried it through the house before leaving it in her bedroom.

Often after a retreat, I walk through the rooms with burning sage, since the herb is supposed to cleanse and bring fresh energy to a space. Recently, I also read that researchers in ethnopharmacology have concluded that certain herbs burned for an hour or more in a room actually provide healthful benefits by killing airborne bacteria. One study indicated that a reduction of 94 percent in airborne bacteria was detected after an hour of smudging, and a long list of pathological bacteria were still absent thirty days later. I hope further testing strengthens these claims; it would be particularly pleasing to know that ancient cultures which made use of herbal smudges weren't just being superstitious. Seems to me that every now and then modern scientists with the latest equipment "discover" something that "primitive" people already knew.

Meanwhile, I will continue to smudge before and after retreats; I enjoy leaving the house smelling wonderful, so if that's the only benefit, I'm content.

September 22. Clouds tumble, the meadowlarks are gone; finches call in mellow notes from sunflower stalks clicking in the wind. Spider webs lace the grass, thick and white, each with a hole in the center leading down to the ground and below, down, perhaps, to the lord of darkness: equinoctial thoughts.

I may not be sure of the precise day of equinoctial balance—it varies according to the cycle of leap years in the Gregorian Calendar—but I feel it in my body when I reach for sweat pants in the morning instead of a skirt, for a turtleneck instead of a t-shirt. I haul in baskets of tomatoes to dry for storage. Like my ancestors, I am particularly grateful for the harvest, for sunlight and bright skies because I know the season of icy barrenness is coming.

Around this time of year I study my house plants on the tables under the deck where they have spent the summer fighting off grasshoppers and outgrowing their pots. I trim, repot if necessary, and bring them inside. I've hung the rosemary from the ceiling in my study where I can nip off a few leaves for garlic-rosemary potatoes. The Swedish ivy and spider plant have already begun reaching for the stained glass butterfly in one window, and the oxalis is blooming again. I've placed big pots of basil and thyme on the low red shelves in front of the basement windows, hoping they will get enough sun. The lavender is beside my computer, where I touch it to release its calming scent when I can't find the right word. My lavender plants always die over the winter, but they provide calm enjoyment on the way to death.

Each autumn I review the summer of retreats, thinking about how to provide more help for writers. I do practical tasks like create new handouts for questions writers ask and mark useful passages in books. I inventory the retreat house supplies to see what needs to be replaced. As I work, I remember each writer and how we worked together and take notes on what I might improve for the next retreat, either in the house or in my approach.

As usual in fall, I'm thinking of ways to make my basement study cozier, brighter, and greener, knowing I'll be spending more time there this winter. Windbreak House Retreats are now open year around, since I am living next door. To encourage winter visits of writers, we even have a blizzard policy: if a writer gets snowed in and can't leave, there's no charge until the highways are open again. Several writers have been delighted to remain snugly ensconced in the retreat house, considering a blizzard the perfect excuse to read some of the hundreds of books available as well as work on their own writing.

Still, fewer people schedule their retreats in late fall and winter, so equinox is the time when I often shift my attention from other writers toward my own writing. I am always aware of whatever book I'm writing. Annie Dillard says that if you leave a piece of writing alone for a week, it may turn so feral you have to approach it with a whip and a chair. After a summer of only intermittent work with the new book, I can almost hear it growling.

As my mind turns toward writing, the sharp fall air stimulates my hunger, so I'm often torn between cooking and writing. A pot of tomato sauce simmers on the back of the stove. My final retreat guest of September brought me a bushel of Washington apples, so an apple crisp bakes in the oven.

Today, I picked what may be the last bouquet of black-eyed Susans. I think about the new book while I make green chile flavored with jalapenos harvested from my garden as well as some extra hot roasted chiles we found at a farmers market last week.

September 23. The full moon is a deep reddish gold rising over the low hills to the east. Pasted in the front of my dad's journal that began in September 1984 is this quotation:

> As for man
> His days are as grass:
> As a flower of the field,
> So he flourisheth.
>
> For the wind passeth over it,
> And it is gone;
> And the place thereof
> Shall know it no more.
>
> PSALM 103:15-16

He was 75; he'd had several strokes that neither of my parents had mentioned. I have read through his journals from then until 1992, the year he died, seeing the evidence of his deterioration on every page, but I didn't see it as clearly in the living man.

How appropriate to my father, whose livelihood depended on grassland: "His days are as grass." To the psalmist, that line apparently meant that the life of grass was brief, whereas I know that grass can be almost forever. Grasses began to evolve seventy or eighty million years ago, encouraging the animals that lived on them. The combination, along with fire that renewed the grasslands, created a culture so successful that grasslands now cover one quarter of the earth's surface, supporting much of life as we know it. Yet the natural grasslands in this country have "all but vanished," say the experts.

Ironically, after a millennia of evolution, many prairies were transformed in a half century by settlers who destroyed native plants in favor of cultivated ones, believing that "rain follows the plow" and other slogans ignorantly conceived to turn every landscape into a farm. Modern range management has demonstrated that, to quote one expert, "grass evolved to be eaten." Grasslands and those who eat grass evolved together, because grazing is fundamental to the food chain. Grass is renewable, grows without mechanical aid if it receives a few inches of rain and sunlight, and needs to be harvested. The rancher's job is just to take care that the process can proceed with as little disturbance as possible. Fences are necessary to keep each family's grazing critters from getting mixed up with the neighbors'. Providing water for the animals may take some human intervention in the form of drilled wells and dugouts. Predators are a natural part of the process, and most ranchers leave them alone. Here on the prairie, we don't usually have to deal with predators that are at the top of the food chain, like grizzly bears and wolves; fortunately for us, the pioneers and emigrants killed most of them before ranchers arrived. I've read the histories; a grizzly bear ripped Jim Bridger's ear off not far from where my uncle George ranches, and wolves were part of the history of this county. Will climate change bring these historical residents back to the neighborhood?

Meanwhile, my father's life continues only in my memory, in his journals—and in the grass he protected on the ranch he once owned.

September 24. Sliced and dried, tomatoes look like round stained glass windows, deep red.

Walking the dogs, we see the end of stories and guess at the action. Under a cedar tree is a pile of coyote scat that looks as if it's wrapped in gray fur. Under another cedar is a great horned owl pellet with a tiny leg bone protruding from it. Beside a lilac bush are several large white and gray feathers. Each tells of a death: the coyote ate a rabbit,or perhaps mice. The owl ate mice, but possibly also a pigeon.

Thinking of my father so much this summer, I've begun to realize how little I really know of his life before he married my mother when he was 43 years old. Most of what I know comes from his obituary, which means my mother knew. He was born at the ranch near Hermosa in 1909, attended grade school in Hermosa, and graduated from Rapid City High School.

A few months ago, my cousin Charlie, son of my dad's oldest sister Hazel, gave me my father's high school ring. I had to laugh; my father wouldn't let me get a ring because, he said, "In ten years you won't know where it is." I wonder how his sister got his ring and if he knew she had it.

His death notice said that he attended for one year the South Dakota School of Mines and Technology in Rapid City, twenty miles from the ranch, then received a degree in Business Administration from the University of Nebraska in Lincoln. He worked in the tax department in Pierre, South Dakota, until World War II, and then went to California to work in the defense industry. His brother, Carl, an Army Air Force staff sergeant, was killed on a bombing mission in southern Europe and was one of the first seven men buried at the National Cemetery in Sturgis, South Dakota. Immediately John returned to help his remaining brother, Harold, operate the ranch.

He told us he was "sent to school" because his left arm was shortened by a siege of rheumatic fever, and the family didn't believe he could do ranch work. He was always skeletally thin. I've found no diploma from the University at Lincoln and few documents to trace his life. A wedding announcement from a friend was addressed to him in Hermosa in 1933, when he would have been 24 years old. A letter from January 1936, indicates he was at 1423 Q Street in Lincoln. I enter the address in an Internet map site, thinking of the old boarding house where I once lived in Lincoln, and hope to see a place where he once lived, but an Arby's occupies the site. I find more addresses and look for them all: one is in the middle of an industrial park, but others allow me to view the houses as though I were standing on the street. I put in the address 746 Rose Street, where he was in May 1936, and am excited to see a two-and-a-half-story brick house with chairs on a deep front porch where I can picture my father studying. Then I realize I am seeing 745 Rose Street, and turn the camera around: nothing but a vacant lot with a sidewalk leading into the weeds.

I lived in Lincoln for a while, but I never recall him saying, "When I lived there. . . ." Did I just forget, or did he never mention it? Now that I think of it, I don't recall him ever saying, "When I was in college. . . ." Did he hate his time away from the ranch? Yet from the time he entered our lives, when I was nine years old, he insisted I would go to college.

I stare at my computer screen, a tool my father never had, looking at the picture of the vacant lot, the sidewalk leading up to air that once held a house where perhaps my father lived. I lean close to the screen, as if I might look closely enough to see him, bespectacled, skinny, carrying a stack of books, slouching up the steps to some room thick with the humid heat of Lincoln in the summer. The vacant lot is a slice of land where he once was.

I can picture my mother at that age more clearly because I've seen photographs. With red lipstick and short skirts, she went confidently out into the world of secretarial school in a little Nebraska town, a hundred miles west of Lincoln.

September 25. I'm drying parsley and basil on the food dryer George built for me. This simple machine is just a box on legs with screened trays that slide out. At the bottom are four light bulbs with separate switches. A thermometer on the top shelf tells me the temperature, adjustable by turning the light bulbs on and off. Even on warm days, tomatoes usually take four or five days to dry completely. Then I pack a couple of handfuls into a paper bag, fold and label the bags, and tuck them into a tight tin box. I keep the packages small because if the tomatoes aren't fully dried, they'll mold and turn black. With smaller packages, even if I've missed some moisture, I'll lose fewer tomatoes.

At 6:45 in the evening, big clouds are massing to the northwest. I can barely see them from the kitchen window but I can feel them through the fabric of the house, moving overhead quickly, building their fury as I read in my easy chair facing east. Finally a trailing gray finger of cloud appears in the east windows, pointing south, with a drapery of rain trailing behind it.

Then I hear it on the new metal roof: a gentle tapping at first, as if a flock of sparrows was dancing; then a thump or two, and finally a drumming as rain sluices off the roof and waterfalls off the deck.

September 26. At 3:15 every morning, a north wind arises and slams the bathroom door unless we remember to prop it open.

About the time we finish walking the dogs this morning, clouds mass in the north and then glide lower; fog oozes down our valley, muffling the sound of the highway. We hear no birds. Our words seem to vanish into the dense air around us as soon as we speak them. The fog hangs like moss in tree branches, and we feel its weight on our shoulders. Jerry goes to his shop, but I want a break from the computer so I start putting away my summer t-shirts and short-sleeved silk shirts and getting out the fall clothes. I spend a dreamy hour making a pile of things I haven't worn in a year to take to the second-hand store in town. My rule is to give away at least one thing for each purchase.

As I work, I think of my readings in my parents' journals and letters during the past few months. Their stories go on in the mysteries they left me. The end of their lives was the beginning of my attempt to form a clearer understanding of who they were and why they behaved as they did.

Outside the window the fog presses thick against the glass. A metaphor: that's how my parents' lives appear to me, muffled and vague.

Sorting through clothes reminds me that my mother shopped at second-hand stores for me but never took me along. She never bought second-hand clothes for herself. She brought me clothes that were often ridiculously out of style and smelled musty and sweaty: journals of others' lives. I hated the hand-me-downs, but perhaps my parents' finances were too tight for me to have the kinds of clothes I envied.

High school. I did not imagine then what my life would be like now. My plans involved the usual: graduation from high school and college into some fulfilling job (war correspondent, daring journalist) while I wrote best-selling novels at night and raised three or four children on the ranch.

When I look back over the twists and turns and tumbles of my life, I see no point in wondering what I might have done differently or trying to picture what my life would be if I hadn't gotten tired of forgiving my first husband, for example, or if we'd had children. When I look at myself, I seem to see the same person I've known all my life, but I know I've gone through profound alterations.

One tiny example of change, I realize as I put a skirt on the "donate" pile, is that I now buy clothing almost exclusively at second-hand stores. Jerry occasionally says mildly, "You know, you could buy that new."

With a sense of satisfaction, I fold the discarded clothes and pack them into a bag for our next trip to town. Ah: the journal of clothes. Looking at the clothes I've worn would show a different woman than the one in my journals.

September 27. Wendell Berry says in *The Way of Ignorance* (which sounds like a contradiction but isn't) that he is sometimes called a "simple farmer," but that he left the simple life behind when he left New York. If we place a value on community life, we should not invest in practices that destroy the health of our earth. Nothing flourishes in our present economy, Berry says, but selfishness, so that a corporation may be polluting while its leaders speak of healthy children and the purity of nature.

The truth of these statements seems to me self-evident. Yet every day more farm and ranch land is paved for supermarkets filled with shoddy goods made by poorly-paid workers in countries whose leaders hate us. More land is subdivided and covered with huge houses whose owners will get up every morning in the dark to drive to jobs in a distant town. Giant companies whose only interest is profit control more and more of our land, shoving aside the people who have worked on it for generations and know its qualities and its limitations.

I would not always have agreed with Berry. I thought being a writer meant being furious when I wasted time on domestic tasks like sweeping the floor and cooking. Gradually I changed my mind about what is important, and began to question the haste that makes waste, whether it's in thinking, in cleaning house, or in writing a paragraph. I know more now about patience, though I sometimes forget to practice it. Moreover, domestic chores often furnish the time for thought that feeds my writing.

September 28. I sliced and put in the dryer a dozen zucchini a neighbor gave me, the major hazard to living among gardeners this time of year. My friend Margaret and I used to put vegetable spaghetti (spaghetti squash) in each other's mailboxes; they grow nearly as well as summer squash. I sprinkled the sliced squash with salt, pepper, and cayenne before putting them on the dryer; they'll be tasty zucchini chips. Also made tomato sauce after collecting a bushel of tomatoes, and in the evening, made yogurt.

Sitting on the deck this afternoon, warm in the sun, thinking of the end of September weather, I copied another paragraph from Wendell Berry.

"Husbandry is the name of all the practices that sustain life by connecting us conservingly to our places and our world; it is the art of keeping tied all the strands in the living network that sustains us."

"Conservingly." Intriguing word. I am conserving by using my gift zucchini, even though Jerry doesn't especially like it; I've shredded and frozen some, and we are nibbling at the zucchini chips. I didn't choose to grow the vegetable, but we won't waste it.

With husbandry on my mind I surveyed our pantry this afternoon, making lists of food I'll collect from various sources before winter sets in. Unlike our ancestors, we can get to the grocery store pretty often and find an amazing variety of foods there, but I like to have a solid stock of stored food ready for familiar recipes or long blizzards. Sometimes I can almost feel my grandmother patting me on the back for being prepared.

September 29. Garrison Keillor read my poem "Clara: In the Post Office" on Writer's Almanac. I missed it, of course, working in the garden, but several people called or emailed to tell me. I'm especially pleased because Clara is a composite character, created from my visits to the Hermosa post office back when it was run by Bud Preston, who had enough space for a small table and a coffee pot by the front window. People getting their mail stood and visited, swapping stories, and one might hear a speech like I put in the mouth of this woman who was composed in my mind from several local women.

Now so many people have moved into the area that more banks of mailboxes fill all the available wall space, and there's no room for a table or coffee. People still visit as much as they can; our current postmistress is friendly and knowledgeable, asks about kids and health and dogs. There's a bulletin board that lists the meals at the senior citizens' center and other news of interest—but there isn't much room for visiting.

Not long ago, in the post office, I met a friend who ranches with buffalo. I told him I'd heard that in an particularly bad storm he'd lost several buffalo, but out of politeness did not ask him how many. He said with dismay that he'd been relaxed inside during the storm, never believing it could be severe enough to kill bison, surely one of the most adaptable of plains animals. Before we could say more, a man neither of us knew moved between us and began talking about how terrible things had been in town during the storm. We both turned away from him, moving closer together, but he stuck his face between us again and again until we nodded goodbye at each other and left. Even though he lives in a little rural town like Hermosa, apparently, that man has no concept of how to join a conversation, let alone how severe the weather can be on the prairie, or what might happen to his grocery budget because tens of thousands of cattle and buffalo died in this storm.

September 30. I arranged Mother's journals in chronological order some time ago, but until today, the anniversary of her death. I hadn't looked at her last journal or her Bible. I'm puzzled to see that she got the Bible from someone in Pharr, Texas, in 1976. Did she trade Bibles with someone? Why Pharr? I see it's near McAllen, where they went on vacation for what I think is the perfectly odd reason that she lived there with my biological father. How did she tell John, her third husband, that she'd like to vacation where she once lived with her second husband?

The Bible is as battered as a good Christian would want, obviously used often, but includes information that wasn't in the Bible she used most of her life. One section is called "What the Bible is About," with a guide to Biblical comments about topics like race relations, right and wrong, and the "mind-twisters," alcohol and drugs. Like all her books, it's stuffed with clippings and letters, including a picture of a little blonde girl in a pink dress on a green lawn, just as she preferred to see me.

The Bible reminds me that Mother became much more blatantly Christian as she aged, until I had to intercept mail and supervise her checkbook. Her nursing home room was stuffed with personal letters and large photographs of crusading evangelists thanking her for generous donations.

With the Bible is her final journal. In it I find one of my last postcards to her, written on September 11, 2001. On vacation in Manzanita, Oregon, we were staying in a house with no television; we knew about the twin towers attacks, but didn't have to watch the explosions over and over. I wasn't sure how we were getting home, since all flights were suspended, but I promised we'd be back soon. I was talking regularly with my assistant, Tamara, who visited Mother several times a week, drove her to appointments, and generally acted as my stand-in while I lived in Cheyenne. Mother often mentioned that Tam was her "favorite daughter."

In the journal, Mother's handwriting is nearly illegible; the last date I can read is September 14. When I got home, I had several talks with her and her doctor because she was refusing dental care even though the bridge holding her front teeth wouldn't stay in her mouth. For two years she had been fighting the repair. The doctor and I both told her bluntly that unless the teeth were repaired, she would starve to death. She weighed only 75 pounds. She repeated that she would die with "my teeth." They weren't her teeth and hadn't been for years.

Would having Mother declared incompetent help? I asked her attorney and her doctor.

"I can't honestly say she is incompetent," the doctor said. "For two years she's insisted she'd die with her own teeth."

The attorney pointed out that having her declared incompetent would be expensive and might use money that she later needed for care.

She was wrong but consistent; all her life, once she had made up her mind, she refused to change it even if she was proven wrong. I could not honestly say she had no right to lie to herself. She liked to quote a Kenny Rogers song, "Know when to hold 'em and know when to fold 'em."

Once during the last week of her life, I asked her what kind of coffin she wanted, intending to shock her into changing her mind about her teeth. "I just want to nap," she said, and turned away from me. Perhaps she'd decided to fold 'em.

I arranged for hospice care at the nursing home and went back to Cheyenne, trying not to feel guilty while writing furiously in my journal about her stubbornness. I made notes about her funeral while I packed to return to South Dakota.

Mother died on September 30, 2001, before I got back. In her Bible I found a postcard she'd written me on that date in 1999, reporting with glee that clever Tam took her to Devils Tower, packing a lunch of deviled ham and deviled eggs!

One of my cousins, with whom I had not been particularly close, came to the visitation at the funeral home and stayed by my side, gently reminding me of names I was forgetting, patting me on the back, and making me rest. She knew what I needed better than I did, but when she died a few years later I wasn't able to be with her.

At Mother's funeral, I reminded my aunt Anne, my father's youngest sister, that my folks had chosen her as my godmother. She'd forgotten; but she tucked a stray lock of my hair behind my ear, an automatically motherly gesture.

Two ladies came into the visitation, looked at Mother, chatted, and then said to me, "We didn't know her, but she looks so nice." Mother would have been pleased and pretended that she knew them.

October

October 1. I scanned my 2001 journal yesterday, remembering that in December, after Mother's death, I often saw things she'd like. Then I'd remember that I didn't need to buy gifts for her.

I also remembered how much trouble her determination to have her own way caused. One winter, on the day my parents planned to leave for Texas, a blizzard was raging. My father wanted to wait until the storm was over, but Mother had a hysteric fit, so George and I towed their car to the highway with the four-wheel drive truck, then followed them to Hot Springs to be sure they made it that far. The back seat of the car was piled up to the windows with her clothes, all on hangers, lying on the seat so they wouldn't get wrinkled.

We all had lunch in Hot Springs. My father said again they needed to go home and wait for a better day. Mother ran around the café talking to truckers, asking them if the highway got better going south. They all said no, that the weather was worse. She screamed that they were all lying to her and shrieked at father to hurry up so they could get on the road.

This time, he refused to listen and drove to a motel where they spent the night. George and I had a terrible time getting back to the ranch through the blowing and drifting snow. As I recall, my parents didn't get very far the second night either, detouring east into Nebraska; their trip to Texas took several extra days and involved dangerous driving conditions. My father called every night, furious.

October 2. Wasps and stinkbugs everywhere; I froze ten pints of peppers.

I have done all this digging in my mother's past because I want to understand her so perhaps I can wholeheartedly forgive her, lay her to rest more completely than I did when the funeral home tucked her into her coffin. For two days I've been ransacking my files, trying to find a copy of her funeral service and finally located it in my journal, of course. Instead of a Bible verse on the front, I included a poem she often quoted, "Crossing the Bar" by Tennyson.

I spent money from her nursing home account on her funeral, trying to recall everything she'd ever said about dying, a topic she preferred not to discuss unless she was threatening to kill herself. She wanted the coffin open and "plenty of makeup" to make her look what she conceived to be her best. She loved pink and roses so I splurged in ways I would normally not have approved. My photographs show the pink coffin blanketed in pink roses. A woman whose voice was not as good as Mother's sang her favorite hymns: "How Great Thou Art," "Rock of Ages," and "Amazing Grace." I could picture Mother wincing.

October 3. Woke up with this going through my brain:

"Sometimes I think
Life is just a rodeo.
The trick is to ride
And make it to the bell."

How true: just ride and hold on until the bell rings. Maybe all this introspection and thinking about the past is foolish.

Singing. We should keep singing.

This was a four-frying-pan morning, as Jerry heated up all kinds of cast iron to make scrambled eggs for both of us with two kinds of bacon. He likes pepper bacon and I had regular; he puts a lot of Whoop-Ass spice in his eggs and I prefer a little cheese. Fortunately, we each came to this relationship fully stocked with cast iron. I'd used it in buckskinning camps, of course, but also heartily believe in it for everyday cooking use. Jerry has been cooking for himself and camping for years, so our only problem was whose pans would fit into the kitchen. All of these humble cooking utensils hold memories.

Today is my mother's birthday; she was buried on this date in 2001. And today I realized that I especially love books where a woman character has a great relationship with her mother; our discord is the reason.

October 4. For the past few days we've been watching a pronghorn buck and eleven does as they circle the house as a single organism, grazing, lying down, and leaping up to run. This is the rutting season, so the buck spends a lot of time circling his harem as they graze and rest. Every now and then a doe tries to trot away, and he throws up his head and chases her until she runs back to the group.

Today when we looked out the window at sunrise, we realized that the harem had increased to nineteen does. About midmorning, another pronghorn buck appeared on the skyline, strolled down the hill, and then suddenly trotted up on the top of the dam below the house. From there he was able to look down on the resident buck and the girls. He appears to be a more mature buck than the first one, with a heavy neck and a curl to his black horns. We could hear his breathy snort.

The resident buck looked up, then stood and faced his challenger. Almost immediately another antelope doe raced away, and the resident buck spun around to chase her. She ran toward the east and uphill, ducking under the fence with the resident buck panting behind. Then the first buck turned and saw that the new buck was now between him and his harem.

The new buck trotted toward the does and the resident buck galloped back, dodged around the newcomer and stopped, snorting and pawing. The interloper whirled away and fled to the top of the dam.

For the rest of the day, the nervous resident buck circled his ladies, keeping them in a loose circle, while the other buck alternately paced back and forth on the dam or lay down and stared at the circle. When we studied the situation just before dark, it was the same.

October 5. This morning we saw only about a dozen pronghorn and the young buck we'd been watching. Perhaps the big male with the heavy horns sweet-talked some females into slipping away with him in the darkness. The animals begin drifting southeast, away from the wind, down the draw and into the sheltering willow trees.

About midmorning, I saw a round cloud hanging in the northwest. As I watched, white tentacles began to drop down from it, writhing, twisting: winter reaching toward us. A cold breeze dropped the temperature five degrees in a few moments, and the wind blew hard, carrying splinters of ice.

Music sang in my head again when I woke up today. This time it was, "Looking out at the road rushing under my wheels. . ." and it made me wonder about the effect music has had on my life, even though I've never been able to participate in it fully. Despite my love of music, I can't make it myself, and rarely play all the CDs and tapes I own. My excuse is that I can't play music with lyrics while I'm writing; the words distract me.

How might my life have been different if I'd practiced music as faithfully as I practiced writing? Did I refuse to work at music simply because my mother wanted me to? Would I have disappointed her musically as I did in so much else?

Oops. I don't want to slip into regret or blame. Reincarnation would come in handy so that I could become a musician and follow that path to enlightenment.

> "I look around for the friends that I used to turn to . . . .
> Looking into their eyes I see them running too. . . .
> running blind.
> No matter what we choose, we're running blind."

Still, I use music the way some people use alcohol, coffee, or medicine: to induce a particular mood or to calm my brain when it flutters wildly from one idea to another. On long drives I've always been able to transform my attitude and shift my thoughts by playing different music, not just to keep myself awake but to get my mind off those thoughts that spiral downward into depression or outward into regret.

Though I can't listen to lyrics when I'm writing, I listen when I'm driving and sing along. When I know the words, it's usually because the song was significant to me at a particular time in my life. This morning it was:

> "Looking back at the years gone by like so many summer fields
> In sixty-five I was seventeen......"

Well, there's part of the problem. In sixty-five I was twenty-two and about to make a huge mistake in my first marriage. By the time I heard "Running on Empty" in 1977 I was thirty-four, divorced, and feeling very empty indeed.

October 6. Today, Tam and I drove to Spearfish with a picnic lunch which we ate at Roughlock Falls, one of my favorite spots in the Hills. When George and I first moved to Spearfish, we often walked behind the falls to look at the world through the pouring water. Because of increased crowds, officials have blocked access to the falls, creating a new sidewalk that allows a view but not access. The reasoning is that people might fall, or damage the walls, or drive off the water ouzels (also called American dippers), small, shy birds who can walk under pouring water. Canyon wrens also live among canyons and cliffs and sing a liquid song of notes that cascade down the scale. I was disappointed not to hear or see either ouzels or wrens today and hope the crowds haven't already caused them to vanish.

I understand why the falls had to be protected; limiting visits to people who wouldn't fall or who would not take responsibility if they did, wasn't possible. Instead, apparently we now must make every place safe for the infirm, protect it from the vandals, and create barriers to keep the stupid at bay.

My reason for being in the northern Hills was to give a speech to a group of ranch and farm women at the new Latchstring Inn. I loved the old Latchstring, with its uneven floors and rickety walls, but I'm told that when it was torn down, rats the size of my dogs came out of the wreckage. The new Inn copies the style of the old and is fresh and lovely.

The women I spoke to were young and old, farmers, ranchers, city-girls-who-married-farmers, all chattering away, dressed in a half-dozen versions of whatever they considered appropriate for a "day out." Some wore nylons and high heels, others cowboy boots and silver buckles; shirts with fringe and blouses with embroidery; hair styled yesterday and hair blown by the wind. Every one of them was interested, alert, smiling and laughing with the other women. Listening to them gave me confidence in the future of rural people.

October 7. I was still thinking vaguely about the influence of music, or lack of it, in my life today as I drove to Hermosa for the mail. The song "Forever Young" by Rod Stewart came on the radio, and in seconds I was crying so hard I had to pull to the side of the road until the song was over.

I remembered why the song affected me so much, but an online search reminded me that it was released in 1988, so I heard it over and over as George and I realized he was dying, and on through his funeral and the months afterward. For the twenty-five years since, I have not been able to hear the song, in any situation, without crying. Sources say it was written for the singer's children and represents a parent or older person singing to a child, but many of the phrases seemed to fit my thoughts about George. I wanted "sunshine and happiness" to surround him since he was "far from home." He had always been "dignified and true," and done unto others more kindly than he'd been done to. He was courageous and brave, and in my heart he would always stay "forever young." His guiding light was strong; he followed the stairway to heaven. When he finally flew away, I was hoping that I served him well.

In my heart he has stayed as the song advises: "forever young, forever young, forever young." When I hear the song I think of his face as I saw it before he died at only 42 years old. During the past twenty-five years, I have mourned him and missed him even as I learned to find joy in my life, and to sincerely love again. The connections in my life have been reduced by death many times since yet strengthened by the love of a good man. As I write these paragraphs, my eyes fill with tears, spilling over onto the keyboard until I have to leave the room.

October 8. Gray clouds roll over one another above us. Long wavering lines of sandhill cranes cry overhead going south in midafternoon.

Last night I gave all three of my saddles—my father's, mine, and George's—to a family in the neighborhood, relatives of Jerry's. We'd talked about this off and on for a couple of months, and finally the day came.

When the older boy saw my father's old-fashioned Duhamel saddle, his face lit up with joy. Suddenly I could see my father on his Tennessee Walking horse, Zarro, waiting patiently for me to climb onto the back of my own horse. I distinctly saw him smile at this long-legged kid, as tall at seventeen as my father ever was. The kid picked up the saddle and headed out the door with no idea that he had been blessed by a ghost.

The younger boy gave his thanks while looking me straight in the eye. When he took my saddle in his arms, the weight nearly felled him, but he straightened and hoisted it over his shoulder.

I dallied inside a few minutes, wiping tears away. When I got outside, the older one was telling the younger that when you put a saddle down, you tip it over, so the horn rests on the ground, to keep from breaking or straining the tree.

They'll probably give George's saddle to their cousins, girls younger than they who are just learning to ride. After they left, I cried from the memories of riding with my father and George, as Jerry held me, but I smiled too, knowing those saddles will be ridden and cared for by other families for generations. Scars in the leather, the journal of the work my family and I did in those saddles, will be overlaid with scars of a new generation.

October 9. This is my grandmother Cora's birthday. The day she died my father woke me at dawn, looked out the window and said, "It's sunrise and it looks like a good day to die." The life force, he said then, is so strong that it never ends, just goes on into something else. Later I wrote a poem about her dying, how she snatched the tubes from her arms. "Rage, rage," wrote Dylan Thomas, "against the dying of the light," but by the time I read Dylan's poem in school, I'd learned the lesson well from my ancestors.

"Forever young," echoes in my mind. "Forever Young." My grandmother was old and forgetful when she died, but her smile in the photographs on my wall reminds me of her patience and love.

Today I realized that giving my saddles away is an admission that I am unlikely to ride a horse again. Of course I didn't ride much for the seventeen years we lived in Cheyenne, but I always had my saddle oiled and ready. I'm sure I'd suffer plenty of muscle pain if I rode again but the worst part would be riding a horse I didn't train. Many of the times I've done that, I've regretted it: no one trained horses the way I learned to do from my gentle father. Horses are intelligent and sensitive, responsive to a caring owner. Too often when I've ridden horses trained by someone else I've been kicked, thrown, rolled on. No, I'm not likely to ride a horse someone else has trained, so I've given up something that was of deep importance to me. The freedom of riding a horse here on the ranch has been unparalleled in my life; the sheer joy of moving in such harmony with a horse's muscles and mind is like nothing else I've ever experienced.

When did I ride for the last time? I can't remember. Of course I would have committed that ride to indelible memory if I'd known it was the last.

October 10. Steady rain all night turned to snow at sunrise. The prairie is tan and white, perfect camouflage for the thirty-one pronghorn that seem to appear in the field as the sun breaks through the overcast. Several bucks with antlers taller than their ears, are so alert they turn when I raise the window shades. Fall is flying and trotting and howling along right on schedule.

My brain continues to set my life to music. "I remember what she said to me," sang Bob Saget in *Against the Wind* hum, hum, oh yes, "Wish I didn't know now what I didn't know then."

That desire not to know now what we didn't know then must be universal. Once we've eaten of the fruit of the tree of knowledge we're stuck with whatever it is we know. The trick, I think, is to let life educate us without becoming so cynical we lose trust in ourselves and others. I was so naïve when I married for the first time I can hardly believe it myself. I'd been raised by honest people so my trust was complete and when it was broken, not just once but repeatedly, I thought I might never recover. For years, I fled from every man I got close to, unable to give my trust. I did some cruel things to perfectly nice men because I believed none of them could be trusted not to break my heart. "Trust, but verify," famously said a president, but with some events, like love, we have no choice but to trust and go ahead.

During this first snow of the season, early and unseasonal as it is, the dogs stay inside unless I shove them out the door, and all day I feel like napping. Some natural principle must tell us to hibernate, calling on impulses buried deep in our brains and sinews. Even though the cold is unseasonable, this false onset of winter must activate the instincts that helped our ancestors, both canine and human, survive.

When I look out my basement study window, I see that snow-covered grass in the dogs' pen is alive with sparrows and other small birds, all scratching and pecking at the seeds and bugs they uncover. Every now and then, several of them dive into the tangle made by the vines growing on the gazebo, where no snow reaches them and surely they are finding weed seeds and other nourishment. Several are perched on the angled supports under the deck, fluffing their feathers and twittering. Their biological clocks are singing to them of winter too.

October 11. In a mystery, I read this line, "the depths hidden in any family story—the small uncorrected lies, the accepted hatreds, the unspoken loves." Yes. So often I find an echo of some important truth even in a book I consider to be recreational reading.

Thoughts of my father always call up his straightforward honesty. These days I find it hard not to think also of his prejudices: his belief that the races who lived in the world's cold north were smarter than those who lived in the warm south, because the northerners had to work harder for a living. I'll grant him that a cross-section was harder to see in American life when he was growing up on a ranch in South Dakota, but he went to a technical school where people from many nations studied. He worked in California during the war, and he lived through the 1980s, when we saw many examples of the achievements of people who weren't white. Still, he never shed that narrow belief. In retrospect, I'm sorry for what he missed.

My mother grew up with the same ideas, which he encouraged. When she was in the nursing home and going deaf, whenever she saw the home's lone black woman resident, she'd lean over and bellow in my ear, "She's BLACK, you know." The first time she did it, I went to the woman and apologized. She said, "I pay her no mind; she does it every day." After that I just smiled at the lady and said to my mother, "She knows that, Mother."

October 12. Now that I've started thinking about music, I keep recalling lines that have been important to me. In 1972 after my first husband and I had moved back to South Dakota, I was teaching at a college in Spearfish. He had gone on the road for the state's arts council as a poetry teacher in the schools. I soon heard rumors that he was again pursuing multiple women. In May of that year I started hearing the Eagles sing "Take It Easy." I was feeling like "Running down the road trying to loosen my load," and I had "a world of trouble on my mind." My mother was going crazy, and I was tempted to run off somewhere and become someone else. I certainly no longer believed that "your sweet love is gonna save me."

Maybe I should have done considerably more "running down the road" and less staying at home being faithful to the past. I bought into an artists' retreat in Vallecitos, New Mexico, a two-story adobe house near the river. I spent happy time writing there and accidentally learning how to conduct retreats. The Spanish language that I'd last heard as a child of four in Texas—not counting, with good reason, the classes I took in high school—began to come back. I could understand conversations at the gas station and the grocery story. I gave considerable thought to moving to the southwest, but in the end, couldn't abandon the ranch.

When the Eagles sang, "We will never be here again," I recognized the truth. Years later, after the retreat was sold, I drove back to look at it and got exactly what a person should expect when they try to recapture the past. The house looked abandoned, the river was sluggish, and dead dogs lay along all the roads. The joyful times I had there were past.

October 13. Today, beside the garage, I picked up a pillow tag with Mother's name penciled in her handwriting; she must have labeled the pillow when she moved into the nursing home. To my knowledge I have none of her pillows here, nor have I moved a pillow anytime lately. I have no idea where this tag has been, but it will do as a message from the dead.

I never found the wedding and engagement rings my father bought Mother. She was wearing them when she moved into the nursing home, but she died wearing a big fake solitaire diamond, discolored and chipped with age. What could she possibly have done with her good rings? And why?

I feel as if I've been in a wrestling match after reading and writing about Mother's last days. Perhaps now I can finally feel as if I've borne enough guilt, analyzed enough, apologized enough, and let her rest. But what shall I do with her journals? If I keep them here, I'll read them over and over again.

October 14. We drove over east with Rick today to look at the east pastures he wants to buy. I hadn't been in some of these pastures for longer than I care to admit. My excuse is that I've been overwhelmed with my father and mother's troubles. I've been right to trust Rick's management, because the grass looks good. His big modern trucks haven't damaged the trails any more than ours did. Rick has filled some trails with dirt and gravel to keep them from eroding, a job an old rancher like my father would never have "taken time for."

What a huge area that is, our pasture and the adjoining ones, with no highways, no housing developments. The only traces of civilization that we can see from there are planes going over. Everything I loved is there: the little spring under the limestone shelf, the tiny dam (very full), the stone Johnny built long before any of us rode these pastures. I kept remembering things: the time George and I went over there to get the cattle after a heavy snow and literally buried the pickup when we drove off a hill that was steeper than we remembered. George was driving fast to power through the snow, and he'd forgotten how the hill dropped off. As the bottom fell out from under us, we both began to laugh and continued laughing hard as snow covered the windshield. Laughing was all we could do. We dug for a long time before we were able to get out of that drift.

Once we were there with George's son Mike, and we sat under a loading chute to wait out a hailstorm, holding the horses, who stood outside with the hail bounding off their heads and our saddles. To warm up and dry out, George built a fire with flint and steel where two walls of a roofless homestead shack came together. We could have ridden home without drying off but he loved showing his son how easy it is to be snug and secure if you are prepared and self-sufficient. I wonder if his son remembers.

October 15. I wake long before sunrise to a shushing sound. Puzzled, I lie in bed trying to interpret it. Finally I breathe deeply enough to scent rain through the open windows: the sound is rain whispering on the metal roofs.

Some of the tomatoes I was trying to ripen in the basement have turned pink, but they are hard and have little flavor so today, muttering "waste not, want not," I put them into a beef stock and composted the vines.

I'm now the older generation on both my mother's and father's side of the family, so my responsibilities are changing. My father and my uncle Harold loved to call me at five in the morning, teasing me if I pretended I'd been awake. They just wanted to visit, sometimes to tell me a story they'd remembered about the old days. They called their sisters too, and certain neighbors, all visiting with one another several times a week though they might only see each other at holidays. Just keeping in touch. Lately, I've been trying to call my cousins more often, for the same reasons. Phone calls: we may not remember long what we discussed, but these are still a kind of journal, keeping connections.

October 16. Frost outlines every grass blade. When the sun rises, crystals disappear at once from east-facing slopes, leaving western slopes still icy and bowls of frost in the low spots.

I'm thinking about my phone calls to relatives. I have no brothers or sisters, only cousins, and some of them are cousins by adoption. Yet we are the closest family any of us have left. I imagine bristle-mustached Swede, Karl Hasselstrom lighting his pipe in the early mornings, firing up the wood stove and sitting down to write in his ledger. Upstairs his wife and five children would have slept, three strong sons and two daughters, enough to populate a township. My father remembered that he used to whistle "Red Wing" while he whittled kindling to start the fire in the stove, a vivid detail my father held for ninety years. Even I have remembered it over sixty years since he told me. I know the song because my mother loved to sing its mournful lyrics, but who can whistle that tune today?

None of Karl's three sons had sons, so I am the only Hasselstrom living here now. I've corresponded with folks who are probably the descendants of Karl Hasselstrom's cousins, but have not really pursued the genealogy. All those potential namesakes have dwindled down to me, and since I have no children I see no reason to create an elaborate family tree.

October 17. I wake thinking of our trip to the Oregon coast to see Jerry's family last year. The night we stayed in The Dalles, looking at the map, I saw a tiny listing for Dee, and remembered that's where my grandfather was killed. That day we'd stopped in Hood River and found the local museum closed, but it had never occurred to me I might find the site of the logging town where my grandmother and grandfather lived. There it was on the map, right where it had always been. In all our previous trips to Oregon, I'd never noticed it.

The next morning we drove up the rocky, narrow road to the townsite, nearly deserted. The town was built in 1906 by the Oregon Lumber Company to serve the families of the mill workers. The first workers lived in boxcars and theirs was known as the "backdoor town" because the houses faced the tracks rather than the street. The mountains in all directions have been cleared of big timber and are covered now with brush. The workers also cleared a plateau known as "Dee Flat," above the railroad tracks where they planted apple and pear orchards which still thrive. At its peak in the 1920s, the town's population may have been 250. Automobiles allowed the workers to live at a greater distance from the smokestacks of the lumber mill, so the remaining housing was dismantled in 1960 when the post office closed. A 1996 fire destroyed most of the buildings and ended lumbering operations.

Standing beside the railroad tracks, I realized my grandmother must have looked at snowy Mount Hood every day as she fixed up a little house for her family. Below that peak, unlike anything she'd ever seen, she bore her first child, my mother, in 1909. She cooked meals, probably fished in the creek, and picked berries. My uncle Bud, her second child, was born in 1912, and they must have felt settled until her husband was suddenly dead in 1914, at twenty-six. I burst into tears, as if the air held the sorrow of that 21-year-old widow of a hundred years ago.

At the depot on that spot, she boarded the train with her two babies; my mother was about five years old and her brother only two. The railroad allowed her to send Elmer's body home to his family in Wheatland, Wyoming, on the train. No mention of embalming appeared on the death certificate. She packed, and probably sold, some of their belongings. The railroad gave her a hundred dollars and a ticket to "wherever she wanted to go." She chose to return to Wheatland.

What were her options for supporting herself and her two children? Many widows were taken in by their husband's relatives but grandmother told my mother that one of her husband's brothers borrowed the hundred dollars and never paid it back. She hadn't finished grade school so she couldn't teach. I believe she worked in a café and cleaned houses, until she and her sister opened the café in Edgemont where she met her second husband.

She never could have imagined that her granddaughter would stand on this spot crying for the losses, her husband, my grandfather.

In the nearby town of Parkdale, I asked questions of a clerk, a logger's mother, who said mournfully that logging had been shut down all summer because of the drought and fire danger. Two weeks before, after several good rains, they started again, but were shut down in a few days because it kept raining and got too wet. Suddenly I was genuinely sympathizing with their hardships, and realized that I am the granddaughter of a logger, a bit of a jolt for my environmental opinions.

October 18. We were watching a movie on Jerry's computer and the therapist in the show said, "We cling to the familiar even when it's bad for us." We laughed and then, "Sugar wafers," said Jerry just as I said, "Circus peanuts." Of course the issue was more serious in the movie—a man who kept choosing women like his mother—but the principle is the same.

Remembering that comment today, I wondered if my almost nonexistent memories of my alcoholic biological father (another unwritten journal) influenced the men I chose. My first husband was a handsome smooth talker, just as my mother described my blood father. I laugh out loud, picturing the two men strolling down the street together, my father's fedora tilted at a rakish angle, and my first husband's poetically long hair blowing stylishly in the wind, nudging each other as they ogle women.

Thinking of the two betrayers together, I find it easier to laugh at what they did to my mother and me. Perhaps there is a special level in hell where they will be introduced to one another and to betrayed women with red-hot pitchforks. Of course I'm assuming I won't be there!

October 19. For days I've been having computer problems. I believed I had set up an automatic backup system but wasn't sure of it, so I backed up the books I've been working on, my poetry, everything, before taking the computer to be repaired. The problems went on and on. After eleven trips to the experts, we finally determined what was wrong and fixed it. I hate not knowing exactly how everything works on the computer and feel stupid because of that lack. Yet every day we drive cars, hiring people we trust to keep them in working order; why should lack of computer knowledge make us feel more inadequate?

During those eleven days I was never entirely sure that my work had been backed up and I had no way of checking. I felt completely numb inside. I could not envision myself rewriting this most recent book, even though it is built around a diary so I do have the original.

Then I remembered walking home with my friend Jimmy from school in Rapid City when we were in, I think, third grade. We took our usual shortcut through the lumberyard, scampering between the tall piles of lumber as the workmen yelled at us, then diving out to the sidewalk through the loose board in the tall windbreak fence.

I remember vividly pushing a rough board out of the way and wiggling through the opening. Looking at city directories, I find two lumberyards between our school and our homes; both were a couple of blocks out of our way, so I'm not sure which one it was.

Our older friends had been learning about fractions, and we were both terrified at the idea of having to learn this new mathematical concept. I recall standing in the sun, leaning on that board fence as we caught our breath, and thinking that I was surely going to die. I could not picture myself knowing fractions. Nothing else had frightened me so much before that day: not the yelling men in the lumberyard, or having to let myself into the house alone when my mother was at work, or knowing I had no father. Fractions! They would kill me.

I'm still not very good at math, but I survived fractions, so I suppose I would have survived rewriting the book. Other writers have.

October 20. Smoke from an "intentional burn" in the hills covers the sun, casts an evil yellow light over the grass and flows up over the rising moon. The local joke is that the Forest Service called them "controlled burns" until too many of them became out-of-control fires. My eyes are burning, my throat sore from the smoke. Helicopters are flying over every few minutes. Is this one out of control?

Thinking about my grandmother's life in Oregon, I remembered one of the things she packed and kept with her for the rest of her life: the baby trunk that her husband Harry Elmer Baker made for his first-born, my mother, Mildred. On April 9, 1953, after I won a grade school contest in oral spelling and got a blue ribbon, my mother showed me the trunk, yet another unwritten journal. *"My Gosh! The treasures it holds!"* I wrote, without describing them.

When I inherited the trunk after Mother's death, I examined those treasures with adult eyes. Mother was born in 1909, so her father Harry Elmer Baker, usually called Elmer, built the trunk then. Five years later my twenty-one-year old grandmother walked into a large room, holding the hand of my five-year-old mother and carrying two-year-old Bud. Elmer's body was laid out in his casket, and she held her children up to say goodbye to their father. "He looked as if he was asleep," she told me sixty years later, with tears in her eyes.

My grandfather's death certificate says he was running ahead of the logging train to set a signal when he slipped and fell under the train. He was beheaded.

The undertaker must, mercifully, have been able to conceal the damage to his body.

Knowing the truth, I open the trunk and realize it smells strongly, comfortingly, of tobacco from the pouch that must have been taken from my grandfather's body by some kind person. Did my mother ever open the trunk just to recall her father? I close my eyes and imagine what it might have been like to sit on his lap while he smoked a pipe.

October 21. The cranes are moving south in long skeins, calls rippling through the broken clouds. Directly over the house they pause to circle, circle, ragged lines breaking and reforming until they are a few hundred feet higher, catching a blast of air heading south, and then they're gone.

I am as ready as I can be for my cataract surgery. I have feared loss of my sight since I got my first pair of glasses when I was nine years old, so I haven't slept well for several days. By 7:30 an anesthesiologist has administered what he calls the "I don't care" drug.

Later, I feel a nudge in my eye and see a vivid square of pink. Then the surgeon is gone. By 9:30 A.M., we're home and I spend the rest of the day continuing not to care. I pick up a mystery, but can't concentrate enough to read. Kathleen Norris once told me that she had the flu so badly that, "Linda, I was too sick to read!" and I couldn't imagine it. Now I simply don't care to read. My eye doesn't hurt but I am happier when it is closed.

October 22. At sunrise I see a layer of smoke lying over the ground everywhere, like fog. Six grouse, chuckling and burbling, hurtle toward the junipers to fatten on berries with the robins and meadowlarks.

My vision is different in each eye, so I lurch around the house, trying to do housework. Then I remember that after my father's cataract operation, he came home wearing an eye patch. He died two decades ago.

Eyes closed, I visualize the eye patch. Black, with black elastic. I wore it once as part of a Halloween costume. Hmm. I open the top drawer of my dresser: the eye patch is there, between the shotgun shells and a jewelry box of my mother's. There is no logical reason I have kept it at my fingertips all these years. With a new strip of elastic, it fits nicely across my left eye. Immediately my balance seems to stabilize. I can walk a straight line. I can go downstairs!

I thank the stern photo of my father above my desk for teaching me to keep things I don't know if I'll need, and apologize for cussing his packrat ways.

October 23. A soft rain began in the night. This morning I woke up thinking about a friend who refused to take as a gift her mother's Fostoria china. Why didn't she take the china, thank her mother, and quietly dispose of it later? I see a parallel here; I've often heard someone complain of being weary of "the same old stories," but later lamenting that they don't know more about their parents. Our parents are always trying to help; youth seems to mean rejecting the lessons, only to have to learn them on our own.

A kestrel has been hunting around the ranch houses and barns. We see it sometimes, stooping with a rush of wings past our heads chasing a robin. Today we stood on the hilltop and watched it hunting. First it zipped past us over the windbreak cedars, sending the robins diving south, toward the lower windbreak. Circling, it suddenly spun toward the cottonwoods at the old ranch house. We lost it behind a plumy spray of gold leaves, but a cluster of birds burst from the leaves of the cottonwood, fleeing down the draw, the kestrel in their wake. Behind another cottonwood it zigged, then flew high and stooped, but pulled up sharply and flew to the dam, where it swooped low over the water. More birds—meadowlarks this time—flushed from the grass and scattered as it swung back over the house, scaring the robins from their hideout in the south windbreak.

My vision is still unclear, but I grab a blanket and settle into a chair on the deck to enjoy late afternoon sun.

October 24. At 4:40 in the morning a glorious full moon is high, glowing silver. Thick black clouds sweep past, lined with silver as is the grass. Cranes twine the sky every day now, whooping and crooning.

What did I do, I wonder, before I began to spend my days reading and writing? Now that I can't do those familiar things, I can't remember. I thumb through my recipe file, finding new tastes to try, and tossing out recipes I've never used.

I begin to concentrate on what I can do: pay more attention to texture, color and scent. In a few minutes' walk outside, I appreciate the vivid red of a leaf from a tiny plum bush, the neon yellow and deep red of a gaillardia in my greenhouse plot, and a furry leaf of mullein.

Recalling Robert Macfarlane's comments in *The Old Ways* about his walks on paths in the United Kingdom and other countries, I'm suddenly aware that most of the paths he walked were made by humans, testifying to human occupation for centuries, modified as their purposes shifted. Formerly the residents walked everywhere, some driving livestock; the author feels connections with walkers through the centuries. Now ancient paths—unwritten journals—fade as people walk less.

MacFarlane walked in several countries, but not America. In the eastern United States, he might have found walkways a couple of hundred years old, but paths in the west are younger. The Cheyenne-to-Deadwood stage and emigrant wagon trains left trails on my land, and our pickups have left faint tracks. The paths left by the Indians who preceded white settlement are mostly speculative. Surely they used the high ridge south of my house, because it provides such a good view of the surrounding countryside, and I have found stone cairns there.

So when I walk through my pastures, few signs of other humans exist. Cattle have followed the same routes to water for years, carving the trails deeply with their hooves. If the water is natural, like a gully or seep, they may be following paths created by bison two centuries ago. We can often detect more subtle trails of deer in the hayfields. On our hillside, the rabbits' furry paws have created corridors through the tall grass. Tunnels chewed and dug under the juniper bushes point at routes twisting through the windbreak trees and all the way to the corrals. The dogs even use these trails as they run to check these hiding places. We've found rabbit fur in some of them, where I suppose a wily coyote followed the trail after a rabbit that wasn't fast enough.

We ramble with the dogs around our hillside, but only in a general way, seldom walking the same route often enough to leave a trail. With the dogs, we walk down the driveway on the gravel. Because we keep the public away, and minimize our own travels to save the grass, we have created few trails even in this place we know so intimately and where we spend so much time.

Yet the Lakota walked and rode horseback here, looking for deer grazing in the willows. The bison scuffed dust as they grazed over the ridge and headed down to the wet places below.

October 25. I am able to move more confidently around the house now, wearing the eye patch for reading, but not when I walk the dogs or play Scrabble. Each day I'm struck by how bright and colorful the world is.

Today when we went to Rapid City, Jerry drove past the nursing home where my mother spent her last years. High-rise assisted living centers stand around the one-story original building, hiding the view she enjoyed of the forested hills.

I always feel as if I should stop in and find my mother in her room. Perhaps if I could sit down and take her hand, we might have a few minutes' conversation that would clear up all our misunderstandings. Perhaps both of us would find peace with the memories. Then the highway swoops away at 75 miles an hour as trucks jockey for the outer lanes and commuters swerve in front of me to take an exit. The highway gives me no time to regret.

I made Swiss steak for lunch just for the pleasure of using the big round black pan with the red lid my mother used; I believe she got it from my grandmother Cora, another kind of journal. As I slide it into the oven, I picture both those women doing just the same thing. My noon meal will be a version of theirs, though neither of them used cayenne pepper.

October 26. I'm still frustrated by being unable to follow my usual schedule of settling at the computer to write, and realize that I sometimes read or write for sixteen hours a day. I need to practice doing other activities—perhaps even doing nothing.

Today when I walk outside, I stand in the sunshine and notice how colorful the grass is on the hillside, blending every shade of red from maroon to pink, segueing into golds and greens. Somewhere on that hillside may be twenty antelope, their tawny sides and white bellies blending perfectly into their surroundings.

Henry David Thoreau says that "two or three hours" walking would carry him to "as strange a country as I expect ever to see." I don't need two hours; walking for three minutes in any direction I can see something I've never properly noticed before.

When writers come to the retreat in warm weather, I often walk with them to some spot where we can both sit and look at whatever is there. I never plan ahead, except in a general way: I'll take them out to the big cottonwood that fell last spring. Then we both sit and write whatever we see in a very small space. They're surprised, of course, because they've never been here before. I'm always astonished too because I've looked at that tree hundreds of times, yet never quite in the same way as on that particular day.

October 27. Today, day six after surgery, I forget the eye patch when we go to the grocery store, so my left eye, with its improved distance vision, is uncovered, and all I can see from my right eye is a blur of moving figures. The visual stimulation is so disorienting I can hardly function. Faces seem to leap at me; colors swirl and shift as I turn my head. Loud music seems to magnify every sensation. I can't read the type on shelf displays very clearly with either eye; letters blur and swirl. I keep reaching up to cover one eye or the other, and can't seem to stabilize myself. I lurch and stagger and catch expressions of pity on several faces: "poor old thing" or maybe "drunk old coot."

How we see ourselves, and how we see, both literally and symbolically, can change so quickly.

October 28. The second surgery takes less time than the first. Again, I can see the tools of surgery and feel them in my eye, but I don't care and nothing itches. Dizzy, I hold Jerry's arm to walk out of the clinic, seeing every detail of faces, of parked cars, and of the men pouring concrete, with both eyes. I haven't seen this well without glasses since 1950. I can't read the paper; my improved vision is for greater distances than my arms reach.

On this date in 1956, when I was thirteen years old, my mother wrote in her journal, *"Today my daughter and I weigh the same, 107 lbs."* I was already taller than her five foot two inches but after this date she never recorded my height again—only my weight, which she always compared to hers:

Oct. 3, 1956 . . .110 pounds (Mother 107)
Jan. 1, 1957 . . . 112
Dec. 1, 1967 . . .117 (Mother 106)

Was she somehow gratified when I began to outweigh her? Eventually, I grew to five feet six and a half inches, so of course I needed to weigh more.

October 29. By the time I try to describe what I saw during cataract surgery, the memory has become too fuzzy to capture in words. At home, I can walk up and down stairs without flinching.

Lately I've been practicing for the changes that will inevitably come by trying to picture someone else living in this house and walking over this hillside. As I water the bushes I've planted around the house, I picture someone eating the fruit from them and wonder if they will be grateful or if only the birds will dine here.

I can't see well enough to get a splinter out of Jerry's finger. I must wait at least a month before having my eyes checked to be fitted for glasses, or not.

October 30. On this second day after the second surgery, I complain that drugstore reading glasses are packaged in threes. "Why do I have to buy three?" I sneer, "Are they supposed to match all my outfits?"

Navigating around the house is easy without my usual glasses, but when I need to read a recipe, play a Scrabble game, read a book, or type, I need help. Soon I have one pair of glasses by my reading chair, another at the computer. The third migrates between the dining room and bedroom. I send mental apologies to the folks who wisely package three pair together.

October 31. From our ancestors came the tradition of celebrating the harvest on this night, understanding that every ending is a beginning. My father's 1982 journal says: *"back on sensible time"* referring to the end of daylight saving time.

Hearing radio ads for Halloween events reminds me that the first Halloween after my mother married John Hasselstrom and we moved to the ranch, I was probably homesick for the town where my mother had taken me trick-or-treating every year. She'd hold my hand as we walked from our tiny house to the high-class section of the city a few blocks away. Then she'd stand in her fur coat at the end of each sidewalk while I paced up to the huge houses—three stories tall—and rang the bell.

My new father took me trick-or-treating in Hermosa. Several neighbor girls went with me, all of us giggling in the back seat while my father drove and Mother rode in front. Snow lay on the ground, and the car heater roared. I felt warm and safe, and as if I was going to find friends, to really belong to this strange place where I'd landed.

My costume was an ugly rubber mask of an old woman with a wart on her nose. It completely covered my head. I could hardly see out the eyes, but people laughed when I looked at them; I enjoyed that. We drove around the six or seven streets of the little town, stopping wherever a house showed a porch light, a clear invitation to little goblins. Always we were invited into the hot kitchens and stood there in our heavy coats while white-haired ladies tried to guess who we were. They never could guess my name, because they hadn't watched me grow up; I still stumbled over saying, "Linda Hasselstrom" instead of "Linda Bovard."

"Oooooooh," they'd say when we unmasked, then nod and smile while they filled our bags with candy.

Finally my father stopped the car in front of the old hotel in Hermosa. The building was two stories tall, completely gray because the paint had worn off long ago. One dim light shone from somewhere inside. There was no light on the porch, and the street light was broken. The place was only a block from a rowdy bar.

Beside me in the back seat, my new friends whispered. One screamed, "She's a witch!"

My father half-turned in the front seat of the Chevy. "I've known her all my life," he said. "Go up and knock on that door." Then his lips got tight in that way that I learned to recognize as meaning he wasn't going to change his mind. I think my mother said timidly, "But if they're afraid of her, John." I don't remember what he said, but we got out of that car.

We bumbled and giggled and stumbled up to the door and knocked. A quavering voice we could barely hear asked, "Who's there?"

"Trick or Treat!" we screamed.

I can still hear the sounds—faltering steps as she moved toward the door. Perhaps the light moved with her, a candle or flashlight, or perhaps my imagination is making the scene spookier than it was.

When she opened the door, with the cavernous dark behind her, I wanted to run. Only the thought of my father's disapproval kept me in place, and the other two were behind me. She urged us inside and led us through what I remember as shadowy piles of things—books, newspapers, chairs piled on top of one another to make a path—to a couch close to a chair with a single light beside it and a book in the seat. There was no TV; my family didn't have one either, but one of my friends was shocked.

The old woman moved books and papers and perhaps a cat off the couch and urged us to sit down, talking all the while about how she'd forgotten about Halloween, hadn't bought any candy.

Still talking, she went off in the darkness and another light came on. We heard water running, and the clink of glass. Cupboards opened; paper bags rustled.

We sat and looked around us. In memory, the darkness towered over us several stories high, but I'm sure that it was only the lobby of the old cowtown hotel. We squinted into the shadows, seeing dusty desks and chairs, then clothes hanging from hooks on the wall. We scooted close together and whispered about ghosts peering out of the shadows, our giggles strangled by fear, conscious of the empty rooms on the floors above us.

What would the old woman do to us? We tried to be terrified, but surely we were all conscious of my parents waiting in the warm car outside, the motor and the radio on, lights off. Why didn't they come in? Perhaps that would have spoiled the fiction of Halloween or forced everyone to acknowledge the mutation of her place in the community.

Finally she came back, still chirping apologies for not really being prepared for Halloween, bringing a tray which she put down on a table in front of the couch. On it were several steaming teacups and a plate of crumbly cookies.

I picked up the teacup, bone china, with a rim of dust she hadn't quite washed away. I'm pleased to remember clearly that I saw the dust, decided it wouldn't kill me and that to wipe it off with my finger would draw her attention and might make her feel as if her hospitality was lacking. I'm grateful to know that I wasn't a complete moron.

The mask had a very small opening for my mouth, so I had to hold the tea close to the rubbery lips to sip.

The old woman cackled, "You look like you've got St. Vitus' Dance." I had to ask my father later what she meant.

We all sipped tea and relaxed while we ate dusty cookies, and she told us stories about living in the hotel. "Sometimes people knock on the door and want a room!" she said, laughing in her crackly old voice.

When she asked who we were, we made her guess. She looked at my all-enveloping rubber mask and said she couldn't possibly figure out who I was. When I told her, she said, "Johnny Hasselstrom's girl, eh?" and I knew in that moment something that had never occurred to me before: that she had known my father as a child and that she might know things about him I would never know. I doubt he and his siblings went trick-or-treating in town, but she may have been a rancher's wife before taking refuge in the hotel. And she did know about my father's late marriage.

I sensed, too, something else, though I couldn't have put it into words until now, as I write about her more than sixty years later. Behind her chatter, behind our silly fears of knocking on her door, hung the whole intricate pattern of the community in which we all lived. Her history hovered over us—perhaps especially over us, children who knew nothing of it—like those dressers and bookcases in the whispery darkness beyond the light. She had once been like us, and someday we, too, could expect to be old.

I don't believe any of us thought coherently enough to fear being old and alone, sitting in a ramshackle hotel in the dark, but we felt it. Perhaps we knew, too, that we might not be able to help ourselves.

Whatever thoughts went through all our minds earlier, by the time we stood up to leave, we were laughing at the stories she told us, perhaps a little in relief. She came with us to the door and dropped into our treat bags a few pieces of dusty, unwrapped chocolate. I think she leaned out and waved to my parents.

Then my memory goes blank.

I lived on or returned often to our ranch six miles south of that town from then until I was nearly fifty years old; I went to church and to community events in Hermosa, but I have no conscious memory of what happened to the hotel or to the old woman. No one is left whom I might ask.

Could I identify her, find out the facts? My memories have restored her for a few minutes, but her life was much more than I know.

In the history of our town published in the 1970s I find no story that I can identify as hers, but one photograph might be the hotel, the Glendale, built by a pioneering town-builder in 1886 and torn down in 1968, about fifteen years after that Halloween night.

November

November 1. Sunrise is the dark horizon with a line of pale green above it, then a line of peach shading into pink just below a dark blue line of clouds.

Somewhere I've read about a man who said he liked to travel to learn about older cultures through his feet, by walking the paths they walked, and seeing something of the sights they saw, even if they had been transformed by the modern world.

When I walk through the ranch yard, I often see it not as it is today but as it was years ago, possibly from an era I dreamed of overnight. Father just steps into the barn, or I wonder if something I'm looking for is in the granary that was once over there. These are not ghosts but shades of memory. Searching family photographs trying to remember exactly where each building stood in the ranch yard isn't much help, because my mother took pictures of cats and flowers. I can glimpse the ranch yard only in the background of a picture.

My nervousness about the eye surgery wasn't really about the surgery, but part of that subliminal fear that haunts us all: how much time do we have? How will we fail—dementia? Disease? I need to laugh more.

November 2. Yesterday I took boxes of letters to the dump, along with hundreds of carbon copies of the letters I wrote in response. They dated from 1987 through about 1998, when I began using email. Scholars are correct in lamenting the loss of history as email replaced handwritten letters.

I had sorted through these boxes one day last summer, seeing single-spaced typed pages filled with intimate details. In some cases, either my letter or theirs bore a full name and address. In others, the person's signature and my response used only a first name, indicating I was writing to someone I knew so well I believed them forever memorable. Now I couldn't identify some of them.

Guiltily, I spent half the night trying to remember names, even searched online. Once I had spent hours of my valuable time responding to the fears and desires and complaints of these people. How could they be gone from my memory? Now, conscious of life dwindling ahead, I work every day at my writing, ferociously battling interruptions.

My choice seemed clear: spend my writing time trying to put these letters in context, re-learning lessons about my own past, or put them out of my mind.

I chose to forget. Besides, those letters belong to the people who wrote them, and neither I nor my heirs could legally publish them. So I tell myself I acted partly to protect those who wrote to me from some future reader who might not know the law or who might be merely curious.

Still it took every ounce of my willpower to heave the boxes into the city dumpster. Today they are buried in the landfill: crumpled pages with crossed-out words, blotted with ink and tears.

Driving away, I felt lighter, but I fully understand that this lightness is partly the loss of those memories, of deliberately destroying some of my own history.

Each of us, I've read, can be summed up by our experiences, but only if we remember them. Conversely, I've read that even events we don't remember imprint themselves on our consciousness. Those events become part of our story even if we cannot define them. I have experienced many things I do not now remember. Yet even lost memories are part of my life. Those memories influence me even if I don't remember the specifics. Either way, I feel stronger for having made a decision.

November 3. The electronic thermometer registered 38 degrees as the overnight low but showed 60 degrees at 4 A.M.! Amazing weather for November, but with tall dry grass around us, we need to be alert for fire. I am astonished at the number of drivers I see on the highway who casually flip a cigarette out an open window on these warm days, with no idea of the potential for catastrophe.

After our noon naps, instead of playing Scrabble we drove into Custer State Park and hiked a couple of miles. Our route took us down into a dry creek bottom full of monstrous boulders and then up through meadows deep in grass surrounding freshly-used buffalo wallows. We watched ahead and behind as the manure became fresher, but didn't see any buffalo. After gaining enough height for good views north and east over the plains, we turned back, having hiked about three miles in a couple of hours. We didn't set any records, but we exercised, enjoyed talking and breathing fresh air, and saw great limestone cliffs and burned snags.

To remember, said St. Augustine, is to re-collect, to sort, and to understand, because *cogo* (collect) is related to *cogito* (re-collect, cogitate, reflect.) Surely that is what I have been doing this year: sorting and, therefore, attempting to understand. I could do so much more.

Returning an email from one of my publishers, I casually told him that I'd thrown away those boxes of letters. He sent me an agitated email, saying he hadn't slept most of the night, thinking about the terrific loss. His distress makes me happier with my choice. He says I should have sent the letters to him for the historical record and his writing about me. Then he'd have been the one to spend his valuable time going through them to determine what, if anything, he had the right to publish.

If we write journals to help refine our thinking, to help us understand life as we live it, then all the confusion in them is simply part of the process. None of us can see into the future, so it's normal to be confused. We live day by day, and each day we see more of how our lives are progressing. When we look back a year later, we might think less of ourselves for not seeing how logically life was unfolding, but it didn't seem logical while we were living it. Perhaps the purpose of a journal has been served once the writer is dead.

On the contrary, perhaps the value of the journals is to remind us how we refined our thinking, how we began to understand the events that were carrying us along. For a writer, then, the journals would serve as a valuable record of how we arrived where we are today, a memory aid providing not only details of the people and events that led us here, but how our minds functioned in delivering us to the present.

November 4. As has been the pattern for a week or more, it was just above freezing at 4 A.M. and up to 66 by noon. Blustery wind. Walking the hillside with the dogs at noon, I collected a massive dry fall bouquet of reddish turkey foot grass and pale gold brome with the seeds intact; tall dried mullein so dark brown it is nearly black; dried brown dock; milkweed pods with fluff still attached; fluffy blonde seed heads of cress; a stalk or two of witchgrass with its wild heads; and one or two tansy mustard seed heads; and a dried aster or two. Clustered together in the cobalt blue jar I found in a second-hand store, these stalks are a striking reminder of what's outside on the hills, waiting for snow or rain to rejuvenate seeds for next year. I collected a few of the invasive species, but on this thickly-grassed hillside, they are few and far between.

The abundance and variety of these grasses remind me that my hillside has not been grazed since we built our house in 1981, allowing it to go wilder than most pastures. We are in one of only a few "temperate grasslands" in the world, the dividing zone between desert and forest. The only other areas that resemble the prairie on all sides of me are the pampas of South America, the veldts of Africa, and the steppes of Eurasia.

This fact staggers me. Local developers sometimes seem intent on covering the prairie with highways, parking lots, and subdivisions as quickly as possible. Yet the Great Plains represents one-quarter of the temperate grasslands on the planet. The plants and animals in all these areas are all similar, and unique to those habitats. All grasslands receive between 10 and 35 inches of precipitation a year. I can picture a wizened African elder recording rainfall just as my father did here, keeping a record for years.

The greatest treasure is the soil: these are the most fertile soils in the world, but what they grow best is native vegetation, not crops that require plowing. Deep grass roots have decayed for generations to produce nutrient-rich earth providing a food source for living grasses. Yet great swaths of grasslands have been destroyed by plowing or development.

What will happen if we destroy the world's grasslands? Who among the developers and pavers and politicians mouthing platitudes is considering the possibility?

November 5. This morning a hard frost covered all the grass with a white rime and now a line of black cows paces in silhouette over the hill toward the tank of water in the corral.

From my journal of 1979 I've recovered another memory I wish I'd left buried. In 1979 my grandmother, worried unnecessarily about expenses, turned her heat so low one cold night that the pipes froze in her little house and her "pretty bird" died. She'd taught the little parakeet to recite the names of all her children and grandchildren, in order. My uncle George was so angry at her, though I understand now that part of his anger was probably guilt at not realizing how troubled she was. Now he is older than she was then and determined to remain in his ranch home in spite of the difficulties he and his wife are having there. I know that I have inherited stubbornness from both sides of the family, and I fear for the folks who be will around when I'm ninety. I'm trying to make plans, though, that will prevent me from being quite as difficult if someone else is responsible for me.

Maybe I've discovered another reason for journal-keeping; maybe we are writing to keep our mental faculties intact. My father, in his last years, took to writing down the names of neighbors, his doctor, the propane deliveryman—people he'd known for years. He was forgetting, and aware of it, and trying to fend it off. Perhaps that's part of this need to write things down: recording the details that make us who we are.

November 6. Yesterday, when the temperature hit almost 65, we rode the four-wheelers in the pastures much of the day. Jerry wanted to learn where the southern boundary of the ranch is, so we headed south along the railroad track, riding slowly through a couple of hundred young cows and a half-dozen Red Angus bulls. We rode slowly across the bank of the big dam, edging past cows. I was surprised to see water glinting in the draw below the dam and to realize that it is not leakage of the impounded water but live springs. We had some terrific rains last spring, but some of these spots had the bumpy surfaces of long-time waterholes. All the way down the draw, too, we could see evidence of floods from last spring's rains: twisted ropes of grass and branches from the few cottonwood trees that have grown in the wet spots.

At the far end of the pasture, we stopped and turned the machines off to appreciate the silence. To the east were the railroad tracks. Recent work has replaced many rails, making me wonder if some new and dangerous cargo will soon pass through this place. Far to the south, dust rose from the nearest road, gravel. On the west, we could glimpse the Black Hills, but we couldn't hear the highway three miles away.

Where we stood was nothing but silence and grass. From a distance the hills rolling away from us looked golden, as if the grass is all one color.

Up close, the colors are dazzling: the little bluestem shades from deep maroon through red to pink, its fluffy seed heads shining in the sun. A grass I couldn't positively identify—its top was nibbled too short—shaded from light green through an almost neon green to yellow. The heads of foxtail barley looked almost lavender in the sunlight; among the colored threads of grass I saw rust and smoke, dusty green and gold, and chocolate brown seed heads.

For a few moments, I wanted to have a horse and ride these pastures, wanted it so much I started to cry and had to turn away. The four-wheelers are efficient and faster, and Jerry would never ride a horse. I miss the quiet, the motion, the muscles, the feeling of the saddle under me. Good and logical reasons exist why I no longer have horses: the pasture is leased and I'd have to take some of it back to feed a horse. I'd have to buy an animal I trusted or spend years training one, then spend several hours a day working with it, time taken away from Jerry, my writing, my dogs, and from my teaching. Still, the desire was strong. Sometimes when I drive past a horseback rider, I have to pull over on the highway and watch, mourning.

After riding the fence line north to the next pasture, we found a gate torn down and a few of Rick's heifers in the neighbor's pasture. Thanks to the wonder of cell phones, I was able to call him immediately and tell him exactly where the damage was. If I'd found a similar problem when I last rode this pasture on horseback, I'd have had to ride home and call him. If he hadn't been home, I couldn't have left him a message because we didn't have voice mail or answering machines.

Then we crossed the tracks and rode through what used to be a busy prairie dog town. The prairie dogs are gone, and the grass is lush where they used to pock the earth with their deep holes.

Riding south again, we reached the southeastern border of the ranch and one of my favorite spots: a jog in the earth where three pine trees stand almost in the fence line along a steep-sided ravine. Far south we could see scattered buffalo belonging to the neighboring Triple 7 Ranch on our south and east border.

On my topographical map of the ranch, I can easily cover these pastures with my outstretched hand. Standing beside that fence my father built, with no sound but the wind shushing through the tall grass, the acreage seems vast. I remember many times when my father and I, or George and I, worked here through the welcome silence and heat of a fall day. No one could call us, and we had no job that day but to fix the fence. We'd talk quietly, laugh, and eat lunch in the pickup's shade. Around us the real life of the grasslands would go on: antelope peeking over a hilltop, birds catching bugs.

One of those days was the last time we would ever enjoy that work, but we didn't suspect it. I hope we enjoyed that last day thoroughly, with no harsh words for each other and with full appreciation of the beauty of this grassland I love.

I've spent several hours thinking and scribbling and trying to put into words what this expanse of grass and these memories mean to me. Perhaps the notes will evolve into a poem.

Sitting in my chair after lunch, I thought about those miles of pasture we had ridden, of the way the grass rustled in the breeze, and began crying. I sobbed for fifteen minutes. I don't believe I cried out of nostalgia; I've mourned my father, mother, and husband, and I understand that my youth is gone. Maybe I cried because I'm in charge now and responsible for what happens to all this land, to all those lives that depend on it, including the badger living on the ridge. I see no point in regretting that I didn't have children, therefore it's not my family who will oversee this land.

Jerry knows my reasons for thinking as I do; he has similar feelings of love for the place where he lived after high school with his grandfather, yet he knows his siblings will sell it when his parents die and understands that he can no longer live there. Neither of us want to run cattle again, so we must live with the consequences of our choices.

I speak as if I really owned the land because a title company searched the records of ownership at great expense and each owner signed documents. While sitting in the company office I remembered the previous white owners as well as the Indians who left little sign of their passage. We whites have left trails, corrals, barbed wire fences, tanks, posts. What will be the signs of the next owners? I hope there will never be a time when the pronghorn are driven out, or the coyotes, deer, and porcupines.

November 7. When I look out the window mid-morning, I see fifteen Canada geese at our dam. Some sun themselves on the bank while others float gently, heads lifted on straight necks like the funnels of warships as they look alertly around.

Two months ago I read an article about drive-lines, the piles of stone prehistoric and early American people used to herd bison toward kill sites. The photographs showed stone mounds like the ones that reach from end to end of the ridge south of our house.

Suddenly I could see those mounds on the ridge south of the house as they might have looked when they were in use: topped with brush or poles so they formed a fence, the bison grumbling as they moved along the ridge, followed by Indians waving hide blankets. The hunters would drive the bison off a bluff, relying on the fall to kill and injure enough so that the tribe could harvest the meat, hides, and bones to feed, clothe, and shelter them through the coming winter.

Experts used to believe that the bison had to be driven off cliffs to kill them, and we have no steep vertical cliffs close to the hill. New evidence, however, suggests that driving the animals off a steep slope might be enough to serve the purpose. On the opposite side of the highway are a couple of sites steep enough to have been kill sites. The picture fills my imagination. I seem to hear those bison murmuring *hunh hunh hunh* as they trot along the ridge.

My questions about inheritance now seem both more complex and simpler. Some legal documents with my signature will eventually determine who owns that south ridge when I die. The only control I can exert over what that person does with the land is to set up obstacles that might preserve that prehistory and the memory of those bison, those primitive herders, whose shadows follow the cattle grazing along that trail on the skyline. We just spent a day with an archaeologist walking the drive-lines, looking at the evidence, and recording it for the state's records. My father wouldn't approve, but he's not in charge.

Walking in the footsteps of those old hunters, I have conjured them up. I have learned about their culture not only through scientific study but through my feet. Unwritten journals.

November 8. Today is Election Day, so I drove to the little white church in Hermosa I attended as a child, where I never go now except on this occasion. Everyone nods and smiles as we walk into the basement, past the church kitchen where several women are tending a coffee pot and crocks that are simmering soup for the workers' lunches.

Behind a table sit several gray-haired women, wearing cheery smiles. They call out our names as we approach. The one with the pen runs her finger down a column until she reaches my name, repeats it, and makes a check mark. One of them hands me a ballot, and I turn toward the flimsy booths opposite their table. A pencil is tied to the counter, and I mark my ballot within a few seconds, then fold it and turn back toward the line of women. The one on the end stands up as I stick my ballot through the narrow slot of the big wooden box with a padlock on the front.

Later in the morning, some of the women will switch places, going into the kitchen to eat the soup they've brought to nourish each other, perhaps with homemade bread. I spot several pies covered with tinfoil on a counter and know they will visit quietly, enjoying each other's cooking as our county expresses itself politically. I picture them, later tonight, drinking coffee as they unlock the ballot boxes and begin to count, but I'm told that they probably send the locked boxes to some central location. There they will likely be counted by officials who didn't nod and smile as each one of us voted today. The totals will be hard to find in the newspaper tomorrow or the next day, but they will be there, proving once again that we live in a democracy. I don't know whether they discuss their own political leanings, but I'm sure they find plenty of other things to talk about, which ought to serve as a lesson to all of us of the many other ways we can be friends without agreeing on how we vote.

We who have voted march outside, wearing the lapel sticker that announces our patriotism, nodding at our neighbors who are coming in—even, or maybe especially, those we know darn well will be voting differently than we did.

November 9. I often wonder what my father thought the future of the ranch would be. If I'd followed standard practice in this neighborhood, I would have married that farmer who was so interested in sticking his bloodline into me while I was still in high school. My father stopped that romance and made sure I went to college. What did he expect me to do? Apparently when he encouraged me to do whatever I wanted, he didn't mean ranching.

What was he working for? He spent his life toiling on this ranch and knew from the time he was in his forties that he wouldn't have a son to "carry on the name," as is the custom. Maybe he was able to put it out of his mind, as I seem unable to do.

November 10. Freezing rain at sunrise turned to four inches of snow.

As I write, both this journal and other writings, I occasionally look back at what I've written. Men seem curiously absent from most of the stories, especially the ones that I've written more than once. Perhaps this is because men are better at keeping their secrets. I have little documentation about my father's work "for the state" as he put it and nothing about his pre-war work in California. I know a lot about the women and have documents to prove some of it, even though my mother seems to have been intent on destroying part of her past. I've read that if you tell men about a problem, they will try to fix it, whereas women will join with you in discussing it and being emotional about it. Perhaps women take seriously problems that men dismiss as unimportant.

Perhaps it's easier to make peace with the women who have inhabited my life because I have evidence, facts, conjectures, and documents. What about the men in the shadows? They are surely not less important because I know less about them, or because they didn't, or perhaps could not, talk about themselves.

The men communicated in different ways. My father proved his love by quitting smoking, by teaching me how to figure out what was right rather than simply dictating rules. He sent me to college instead of letting me marry a man who would have confined my interests to children and milk cows. Would my father say that the documentation I lack is less important than the land?

The women who nurtured me had little ownership in the land, but helped me in other ways. Aunt Anne was the sensible mother I didn't have; Josephine was the tough love mother. My grandmother combined the best of both. But the father who insisted on my education, did not list my mother as a co-owner of the ranch, though he did leave it to her. His only other logical choice would have been to leave it to me and he certainly did not do that.

Maybe the lessons I've been learning in life, and in ranching, and in writing this, all focus on the various ways women learn to survive. My mother probably didn't want a child. Her first two marriages didn't work out, but she tried to do right by John Hasselstrom and worked hard at being a ranch wife. She adapted. Likewise, after my grandmother lost her first love, she found a way to keep going. Maybe that's all any of us do or can do.

November 11. On Veterans' Day we always find ourselves talking about the dead, both the ones we knew and those millions we've never known.

For the past few days I have been doing what my grandmother called "redding up," tidying and organizing several compartments of my life. First I gathered up the muddy, crumpled pieces of paper I'd been stuffing into my gardening journal all summer and read through the summer's notes from planting radishes in the greenhouse through eating the last tomatoes, ripened on the kitchen windowsill. Each entry brought back the tastes of summer.

Since Jerry is retired and I am my own boss, we could harvest most of the vegetables we need from our huge garden plot. As a child and young wife, I considered growing the maximum amount of vegetables part of my job; I dried, canned, and froze food and smugly reflected on the cycle: grow, harvest, eat, recycle, and repeat.

But as my strength dwindles and my age increases, my priority has become writing. So we compromise. I trade beef to neighbors for chickens and eggs, and we grow foods that do well and on which we prefer not to compromise: potatoes, tomatoes, onions, herbs like basil and oregano. We try to grow hot peppers even though buying them would doubtless be cheaper. I still make most of our meals without much help from packaged ingredients, and all this takes a lot of time.

Each fall I draw up a revised garden plan on what I've learned from the summer's work. I sort the seeds I have left and file them in order of the date I'll plant them next year and note what seeds I need to buy. Then I turn to the wild seeds I've collected and dried in the greenhouse, strip the seeds from stems, package and label them. I'll give some seed to the Great Plains Native Plant Society seed exchange and some to friends.

While doing these tasks, I often dash into my office to scribble a note I want to write about, perhaps recalling how the tomato plants looked in July. Simple tasks allow me to think about writing, and I believe the most important part of writing is thinking.

November 12. We disappear, but the land endures. Humans have been in North American for around 40,000 years. The ancestors of the red-tailed hawk perching on the fence post evolved even earlier.

Above my desk is a quotation from Hildegard Flanner, an American poet and essayist: "I stayed at home and cultivated a sense of place and poetry." Friends write to us of travel that sounds fascinating, but staying home, I cultivate my sense of place in my writing.

I'm trying to get over berating myself for *not writing* on those days when I spend time filing or sorting seeds, in part because sometimes the simple tasks lead to writing. After I finished that chore yesterday, I wrote memories of my grandmother to have the pleasure of *having written.*

Perhaps readers picture writers sitting at our computers, writing long seamless sentences that flow smoothly onto the pages of printed books. My writing life alternates chaotic flurries of ideas with flat spaces featureless as bathroom tile. Writing is always, for me, a stop-and-go business. I collect bits of memory and image and dialogue and story and imagination. I stitch this colorful jumble together, pricking my fingers often, into something that I hope resembles a quilt. It may turn out to be a rag rug.

"Redding up," grandmother called it when she put our empty teacups into the dishpan. Then she brushed the gingersnap crumbs off the round oak table (the relative who took it painted it black) into her hand and tossed them out the door where the hens pecking around the rock step would murmur "*Cluuuuuuck*?" and tidy away the morsels.

From the tea kettle she kept steaming on the woodstove, Grandmother would pour hot water over the dishes from lunch, add soap and cool water from a pitcher. She'd wash while I dried them on a soft dishtowel that's likely still in my cupboard. Then she'd hang up her apron, and we'd sit down in our chairs to read until time to feed the chickens and gather the eggs. The only sound would be turning of pages, the *shuuuuuush* as the coals settled in the stove, the tick of cooling cast iron. Having done a little "redding up," we were prepared for whatever came next.

November 13. My memory, another kind of unwritten journal, is never linear the way a real journal would be; my brain never announces, "On September 25, 1951, I saw...." A spider on the basement stairs slaps me back into my four-year-old self, hanging under the porch of the house in Texas looking at the gleam of black widow spiders in their webs overhead while my mother screamed at my biological father in the kitchen above me.

The spider's shiny carapace makes me see what I had forgotten: that while they fought I reached up and knocked a spider onto the sand, turned her over to see the red hourglass on her belly, and poked her until she bit me. I knew what I was doing; my mother was deathly afraid of spiders and had warned me never to touch a black widow. My screams brought both my parents running, and both their arms were around me as they ran for the car. That's what I was after, that comfort of the arms around me, of both my parents taking care of me.

Here's irony. I dumped those boxes of letters from others to me, along with copies of my letters to them. Now a long-time friend who was cleaning out her storage unit in preparation for a move sent me the letters I wrote her between 1970 and 1988. I threw away some of my history and now more has come back to me. We can't escape our past.

November 14. This morning, as we walked the dogs around the retreat house, I could see the yard as I first saw it sixty years ago, with the walkway leading to the outhouse that we no longer had to use after the new house was built with indoor plumbing.

I remembered LeRoy and the kitten. I must have been nine years old, sitting in the afternoon sun on the board walk to the outhouse cradling kittens in my lap, looking around at the fields and barn. Since I'd been able to read, I'd wanted to live in the country, and I was content. My mother had promised me a horse and a father, filling my head with vague images because I didn't remember ever having either one. So far, except for the kittens, I wasn't impressed with the ranch.

John—I didn't call him "dad" for years—wouldn't let me near the horses. His dog, Schnitz, a sausage-shaped mutt covered with a slick mottled coat, was part of the work force, with no time for children or petting. Schnitz helped herd the cattle when we moved them from one pasture to another, his long snout and slim foxy legs never still. He snapped the necks of mice and moles even while at rest, panting in the shade of the truck.

Then one day, chasing rabbits ahead of the mower, Schnitz miscalculated a jump. The blades cut his back legs off.

LeRoy, the hired man, came to the house to confess and borrow the rifle. My father, who'd lived as a bachelor with that dog for five years, just shook his head and said, "Too bad."

LeRoy was subdued at lunch. I tried to cry over Schnitz but we hadn't been close. My father's rule was, "One dog is a dog. Two dogs is half a dog," so I was hoping I could get a dog of my own. If that didn't work, I'd console myself with cats.

So that evening, when LeRoy headed for the house to talk to my father about the next day's work, I was sitting on the walk crooning to kittens. When a yellow fluffy kitten scrambled out to sniff LeRoy's enormous work shoe, he grinned and lowered one foot over the kitten's head. LeRoy's teeth were brown and something black hung at one corner of his mouth.

"Wanta see me squash him?" he said.

I stood up, dumping kittens, and kicked him hard in the knee. He swayed and lost his balance. Crunch. When he lifted his foot, the kitten's eyeballs bulged and its head was flat and trickling blood. "Gosh," LeRoy said, scraping brains on the edge of the walk. "I was just teasin' you."

I kicked him again, bawling so loudly my mother and father rushed out the door. LeRoy stumbled over his feet trying to boot the corpse under the walkway before they saw it.

Howling, I stuffed kittens inside my shirt screaming that I had to save them from LeRoy. When I could see LeRoy's leg through my tears I'd kick him again while spooked kittens yowled and scratched my bare chest so I squalled louder. LeRoy kept yelling, "Ouch! It was an accident. Shit! I didn't mean to. Hell! Damn that little snot can kick!"

My new father grabbed a shovel to scoop up the kitten, snapping, "Don't use that language around my daughter!"

Mother dragged me into the house as kittens scrambled to freedom. Mumbling something, LeRoy staggered off to his pickup.

LeRoy brought me a half-grown brown and tan puppy the next day. I snatched the pup out of his arms, saying, "I'll hold him while you leave, so you won't step on his head."

I named the dog Teddy; he was my faithful companion until I went blithely off to college—forgetting our fellowship was half my responsibility—and broke his canine heart. He died during the first year I was away.

The vision of the walkway brought back the episode with LeRoy, stuck in my brain for nearly fifty years. Transformational grammar, important telephone numbers, and a lot of other information vanished from my mind before it occurred to me that I was at least partly at fault for the kitten's death. LeRoy was big but kind; he would never have hurt the cat or me.

November 15. Outside my window, grass is bent and brown, but on the big south ridge I can imagine the greens and reds of spring.

Never before have I questioned the origin of the expression "redding up," but I investigated online today. Some folks think the expression is southern, but Pennsylvanians use it. One writer says it's used in the Shetland Islands and another quotes a Scots dictionary: "To clear (a space, or a passage) by removal of debris, undergrowth or other encumbrances."

Scots immigrants influenced Southern culture, as anyone who listens to bluegrass music knows. My grandmother Cora was probably born in Swann, Missouri; she moved to Wheatland, Wyoming when she was seventeen, so she likely learned the term from her Southern parents. Smiling as I thought of the way her eyes twinkled behind her glasses as she "redded up" her little cabin, I've bustled around for several days picturing her making tea to sip with cookies. The cookies were homemade until she discovered gingersnaps; she loved to dip them in her tea. The cookie jar looked like a tree stump with a squirrel perched on top for a handle and always sat on the round oak table. I suppose the relative who painted the table black took the cookie jar too.

November 16. The sky is overcast at mid-morning, completely covered with frothy gray clouds. The sun makes a cold blue-white circle high in the sky, just as it looks before a heavy snow. No snow is predicted, but we shiver anyway.

I watch the dam south of the house, hoping Rick's heifers don't try to walk on the ice. We've had some cold nights, but also some warm days, and the ice might not be thick enough. One awful Christmas day, my dad fed a bunch of cows the usual hay and cake in a pasture watered by a similar dam, then chopped a hole in the ice so they could drink. Normally he would have waited to make sure the cattle drank a few at a time and didn't crowd onto the ice. But my mother had warned him not to be late for the family dinner, so he left before the cattle had all drunk. The next day when we went to feed he found a half dozen cows dead, drowned and frozen into the ice. The look on his face, the long silence in the truck as he saw the evidence of how they had struggled to get out—I'll never forget it. He was sorry for the loss of income, of course, but sorrier for the cows' suffering.

Today most of the cows went around the dam, headed for the water tank in the corral, but three or four young ones gathered on a peninsula that juts out into the water, staring across at the cows walking on the other side of the snow-covered ice. I watched them off and on for an hour while we ate breakfast as they, like a group of teenagers, tried to encourage each other to cross on the ice. Finally one walked out a few steps and stood for a while, looking down at the ice as if listening. As she turned to go back, she slipped and scurried back to the bank. I kept stopping my work all morning to look out the window to see if they would cross or fall through. Finally most of the cows had passed by on their way to water, and the remaining cows walked around the dam to follow the others.

This afternoon, I walked the dogs to the west of the house and found two huge holes. I stuck a mullein stalk down one for at least three feet. Cosmo crept down one hole until he completely disappeared; I panicked and called him and he slowly backed out. Either a coyote or a badger might have excavated these holes while trying to get a vole or rabbit. We used to have a badger that would dig a little napping hole up here above the dam and rest in it between hunting for frogs at dusk and then again about three in the morning. Every day we could see a journal of the hunt in its tracks and the marks of hair in the soft earth.

November 17. This morning we saw a big coyote crossing our entrance road, about thirty feet from the house, while we were eating breakfast. When we ran to the deck and yelled, he stopped and looked at us, ducked under the fence and then stopped and looked at us again before trotting off. We backtracked him around the Westies' grave, past the railroad tie fence, and through the windbreak trees north of the house, concluding that he's way too comfortable this close to the house. This is why we usually go outside with the dogs if it's getting dark, or before sunrise, even in our dog pen.

Yesterday afternoon, knowing the cold night was coming, the cows went over the big ridge to the south. This morning they are coming back, a few at a time showing their heads above the skyline, then moving along the ridge and choosing one of its wrinkles to walk down. Long lines of them trace the folds in the hillside, moving down until they strike the east-west fence, then falling into place behind one another, a black bovine column heading east to the gate that will lead them into the corral with water and to the shelter of the snow fence.

As I watch, snow begins gently falling, erasing the view until the cattle are only a gray tracery on the hillside, as if I'm looking at a sketch of the ranch's past. Or perhaps its future when cattle are just a memory here.

November 18. The moon is nearly full. The wind blew some of the snow away today, so in daylight the drifts were interlaced with parcels of brown grass. Tonight, each drift of snow glows in the moonlight like a handful of pearls.

I just sold my father's 1951 and 1962 Chevrolet pickups to an antique auto dealer. Before he loaded them, he gently handed me the Alka Seltzer bottle containing the registration from the glove compartment of the '62. With tears in my eyes I watched the pickup, such a big part of my ranch life, go down the road. Many of my significant conversations with my father took place as we bounced over east in it or dug out of a snowdrift. I taught my first husband's children to drive in that pickup. Now those children don't even send me Christmas cards.

Laughing at myself, I placed the Alka Seltzer bottle on a shelf, as if on an altar. I don't believe any element of my father's spirit hovers around the things he owned, or even the things he loved, let alone a recycled bottle. Still, I feel it's symbolic. He saved that bottle and made a different use of it, his frugality investing it with value. He's gone, but his thrift, and I hope much more of what he valued, lives on in me.

November 19. At a few minutes after 6 A.M., a line of gold splits the sky from the land: black cloud above, black land below. Slowly the gold line expands from the thinness of wire to a broad golden torch, then slowly begins to flow upward, revealing the lacy underside of the black clouds.

Under the juniper trees, where the snow is two or three inches deep, the spiky tracks of sharp-tailed grouse meander among the grass stems. Ahead, somewhere in the tops of the junipers, I hear them chuckling, gurgling, conversing as they eat juniper berries. I've read that when wildlife is stressed in winter, even one wild run or flight might use up enough energy to make them easier prey to cold or predators. Since I want these grouse to come back to eat grasshoppers, I slip away. I've laid chicken wire over my perennials, so they can't scratch up the new plants.

November 20. In our windbreak, I found evidence something has eaten a grouse. The feathers are spread in a circle a couple of feet wide; otherwise only the wings and part of the spine and pelvis are left. I kept a few of the bird's side feathers, decorated with wavy stripes of gold, brown, gray, white and ivory so as to give a mottled appearance so the bird blends with the grass. The head, with its eyebrow ridge of yellow—the only bright spot in the bird's camouflage—was missing.

We noticed big ungulate tracks in the ranch yard today: a single animal. The tracks were large enough to be a small elk, but it was probably a mule deer, a rarity lately. We tracked it through the garden and lost the trail in the grass.

The journals from my mother and father that I have been poring over this year, with the letters I have, the photographs, my journals, all represent three generations on this particular piece of land. This is not much in the history of the world, but a lot for a small fragment of the Great Plains.

What good are my father's records of rainfall and my mother's notes about cleaning house and going to "Ladies Aid"? This is a small chunk of real estate that may be overwhelmed by the demands of a greedy, restless culture while I am still fighting to save a little fragment of grassland. Who cares how my father kept this ranch solvent and the grass and wildlife healthy for sixty years if it is eventually chopped into hobby ranches for incomers driving double-cab pickups and wearing big hats with nothing under them about ranching? Who might care in the future?

I think a lot could be written from these documents, but who would read it? Where shall I put them to preserve them for some future Great Plains historian? Shelves in both my houses are stuffed with such things, and if I don't make decisions and clear those shelves somehow, someone else will. At best that person will be my partner Jerry, who will do it thoughtfully, but I hate for him to have to decide. My cousins are unlikely to value these papers.

So it's up to me; I helped my aunt Josephine decide where to give her father's journals, and the same repository may be interested in those of my mother and father. I'll contact the officials soon. Just deciding makes me feel lighter, as if those dusty boxes were already gone from the sagging shelves.

Once I make a decision, I need to remember not to regret it. I need to do all my pondering before the decision, then let it stand.

November 21. Walking down toward the retreat house every day, we watch the treetops around it and often see hawks or owls take off. Today it was an eagle, rowing the air with great curved wings as it flew south, a slash of golden brown against the frozen blue sky.

For a change of pace from my father's notes, I got out my penciled diary from 1953, when I was ten years old. My first entry was to lament that we'd had two storms before I started writing, *"so I am a bit late."* On Easter, I wrote that I got a dollar as a gift, so I had collected $21.52 toward buying a horse. On May 2, I wrote *"Papa bought me a bridle for $12.99. His wallet looked a lot thinner when he did that."* I must have been getting pretty sure I'd get that horse. For several days after that my daily entry, consuming an entire page, was, *"Nothing happened today."* So much for the myth of the observant writer.

On one of those days, though, my mother wrote, *"The usual Sat. trip to Rapid—dancing class,"* indicating she was reading my diary. I was taking ballet lessons, thanks to my mother's vain hope that I would grow into my big feet and become graceful. I can sum up the results by saying that when most of my friends were being sugar plum fairies in the recital, I was cast as a golliwog.

A few pages later I wrote across one page of the diary in big brown crayon letters: "PLEASE KEEP OUT!" After that I always hid my journals. Even when I was an adult, if we left our house on vacation, she'd insist she had to water my plants, but I could tell she'd gone through my drawers, searching. She never found them.

November 22. This morning before sunrise a column of gold-white light shot straight up. Internet research told me that this light pillar is created by the reflection of light from ice crystals with near horizontal surfaces. When the light comes from the sun, the spectacle is called a "sun pillar" or "solar pillar," but it can come from the moon, or even from streetlights.

The wind has been blowing ferociously all day long; the weather reports say thirty to forty miles an hour. Window screens rippled, and heavy metal deck furniture blew over. Buckets and the empty garbage cans are rolling around the yard. After dark, when the wind has died, coyote songs trickle like moonlight through the clouds.

November 23. When we walk through the ranch yard, we are accompanied by the fragile twittering of sparrows as they zip among the branches of the lilacs and junipers. We almost never get a close look at them; they move too quickly.

Today I found a dead hawk under a juniper tree in the windbreak north of our house. She is tiny, with gold breast feathers, her back and sides mottled brown and white, and broad brown and white stripes cross her tail. One clenched foot still held a curl of brome grass. She was frozen; I gently moved her wings, head, and neck, but could see no obvious reason for her death. I considered giving her back to the sky that is her habitat, but instead called the Game, Fish and Parks Office, took her there to be mounted, to help educate people about these beautiful birds. When I handed her over, the officer who took her told me it was illegal for me to carry her in my car to his office. I hope he was embarrassed to be lecturing someone who was trying to help.

November 24. In late afternoon I settled into a cracked plastic chair against the warm garage wall, facing the late-afternoon sun, and read the drafts of poems I've been working on. Cosmo lay under the chair and Toby hopped into my lap.

At noon, while Jerry was in town, I walked the dogs on the far west side of the hill, where we rarely go. The grass was deep, and Toby was nearly hidden as he followed me. The cedar trees I planted in the deep loam when we built the house have grown extra tall and thick. As Cosmo disappeared into the foliage, I wondered if he'd meet a hidden coyote or badger. Deep under each tree are spots where something has scratched out a resting spot to keep dry even in snow or rain, waiting for prey.

This afternoon, again, I searched dusty boxes of documents and confirmed that I have found almost nothing about my mother's marriage to Robert Osborn or their lives together.

Did she deliberately destroy evidence of that first marriage, perhaps to keep me from learning about it? I realize how easy it would be to destroy a record, especially if there were no children of the marriage.

Finally, I may understand some reasons why she concealed that marriage. The obituary I've written for myself omits mention of my first husband's name, though I intend to name his children because they remain important in my life. I may be emulating my mother, but I have retained all the legal documents relating my first marriage. Surely all her friends knew of her marriage, and she lived not far from where she was reared. Yet I recall her saying at times that she'd been married twice, not three times. Was this a lie, or had she forgotten?

November 25. I enjoyed my favorite kind of awakening: with the lines of a new poem going through my head. I propped my pillows up, grabbed my journal from the head of the bed and immediately began making notes and writing lines as they emerged from my brain, or dream, or wherever they had been lurking.

Slowly the green leaches out of the hollyhocks, some of the last plants to lose their color in winter. The clouds are low and heavy, rolling in a sullen gray stream. The prediction for Thanksgiving is one hundred percent chance of snow.

Besides gratitude for our lives, we're thankful we don't have a TV. As one of my favorite mystery writers says, "What we call news isn't new at all: wars, murders, famines, plagues—death in all its forms." Instead, she says, life is just a "series of overlapping stories about who we are, where we came from, and how we struggle to survive." Surely this is why we identify with novels, TV stories, soap operas and poems: all just different versions of the same story, how we grasp life.

November 26. The forecast of snow was exaggerated, so we gave thanks in the traditional way: by contributing food to a cousins' Thanksgiving. We were gathered in a new house in an exclusive subdivision, but the stories were all family history. One cousin told how her mother, Hazel, got a thousand baby chicks every spring. Local people ordered chickens and eggs from her. In the fall, the family butchered a hundred chickens a day for ten days. My boy cousins kept a fire going under a big tub all day. Hazel whacked off the heads, and Pat scalded each chicken and pulled the feathers off.

"That smell!" she said. Once the feathers were pulled off—gently so as not to tear the skin—the birds were gutted, washed, and the carcasses plunged into cold water, then packed for customers. I helped my grandmother with all these chores, but never more than one or two chickens a day. I'll never get the smell of wet feathers out of my nostrils.

The wife of another cousin, raised in a city, had not known that a chicken had to be scalded. As we described the smell to her, we realized how we'd simply accepted that horrible job as routine. Eyes sparkling, a mischievous cousin mentioned lefsa and lutefisk, and we explained those edibles too, being far too gleeful about her clear distaste for codfish soaked in lye. Though we were all Swedes and Norwegians, none of us can stand the stuff either.

We talked about how it must have been when Karl (he Americanized it to Charles) Hasselstrom courted the widow Ida Callahan, who already had four children and a farm. After Charles and Ida were married, they produced five more children: Hazel, Harold, Carl, John (my father), and Anne. Our dinner was with one of Anne's sons, and a son and daughter of Hazel. They never remind me I don't share their blood; conversely, I'm the closest thing to a family historian we have, a fragile link with the past since I don't really study the genealogy.

Apparently, Harold had a twin sister who died at birth, which reminded me that I remember one of his half-sisters talking about seeing a baby laid out on the kitchen table, being washed and prepared for burial. No baby Hasselstroms are buried in the Hermosa cemetery, and Harold once told me she was buried in the grove behind the ranch house. None of my older cousins had heard the story, so if it's true, I may be the last person who knows it. If my memory is wrong, this may not be history; if I'm right, little trace is likely to be left of the grave after more than a hundred years. Another family mystery.

Many pioneers who lost babies buried them on their own land rather than going to the expense and trouble of a funeral, so burial at home is likely. When I consulted a local historian, she mentioned another family that had "buried a baby in a shoebox somewhere on their place." Beside our trail to the east pastures are the two oval piles of stone I've wondered about for years. They look precisely as graves protected by piles of rock would look, and they are near a tiny dam in a sheltered gully that would have been a perfect early homestead site.

I don't recall ever hearing Harold talk about his sister, but I'll bet he thought about her through the years as he grew up and married—maybe particularly because he never had children of his own. The Hasselstroms have not been good at reproduction.

November 27. With a few calls today, I may have solved the mystery of the burial of Harold's twin sister. A local funeral home that has always handled Hasselstrom burials has a record of a Mary Hasselstrom being buried in the Hermosa cemetery in 1901, though the records don't indicate where, nor is there a headstone. She was probably buried in the plot that became the graves of her parents; perhaps at that time they couldn't afford a headstone, and no one thought to add her name after they died. Of course, it's too late to ask Harold.

I leave my parents' journals on their shelf and reach for a leather bound book just the right size for a man's shirt pocket. Ornately lettered on the front is "Ledger."

Inside the front cover, written with a pencil in elaborate cursive, is "Wm Snable." Bill Snable moved into the neighborhood in 1912 to homestead in the pasture across the railroad tracks a mile from my parents' house. He built a tiny two-story home we can see from our dining room; the first floor walls are made of uneven stones collected in the rocky pasture and the top story is standard beams and planks. He dug a well by hand and kept a few cows. In 1948, he sold his land to the Hasselstrom Brothers, Harold and my father John, and moved to the Veterans Hospital in Hot Springs where he died in 1950.

As a child, I would often ride horseback to his abandoned house and tiptoe inside. The place was small in every way; at eleven, I was taller than the hand-carved mantel on his fireplace. A carved molding at the top of the kitchen walls was painted in vivid red and white geometric designs. His books stood on handmade shelves along with a carton of calling cards printed with his name. I once took home a handful of the cards, and my father was furious; that was stealing. He made me replace the cards. Bill Snable's home was his journal carved out of the prairie.

November 28. The hawks are hard at work early this morning. As we walk the dogs, the kestrel cruises overhead, wings outstretched in a cross, head tilting as it floats, eyeing us. Gently it passes by, gliding over the field until we lose sight of it in the cottonwoods to the east.

As we walk back to the house, we see a red-tailed hawk, wings beating furiously, twisting and turning just above the earth, possibly following the meandering track of a vole below. Then the hawk drops, and its head dips as if it's stabbing prey.

I remember my father and Harold talking about Bill Snable walking past their place when he went five miles to Hermosa for supplies. Toward evening he'd walk back with his groceries and often stop for coffee or a meal. No doubt he kept the notebook to mark his progress in building, writing in pencil, *"commenced work on Bunglalo Sept. 24, 1912."*

halled stone to build
Bungalo, with wheel-
barrow Sept. 24
" " 25
" " 26
" " 27
hired team of Mr.
Jno Crawford to (28)
hall cement from
Hermosa to my claim
dug and wheeled stone
to day 29
" " 30
dug trench for fou-
ndation of Bungalo
today Oct. 1

Picture that man trundling his wheelbarrow over the shortgrass prairie, hummocky with rocks and badger holes, collecting tons of stone. In six days he had collected enough for the cabin's first story, and plenty are left. If he stopped to rest, he could look west down a long sweep of prairie toward the Black Hills, faintly blue from that distance. Horse-drawn wagons and a few cars moved north and south along the Cheyenne-to-Deadwood trail on the route that is now Highway 79.

Day after day, he worked, laying stone, trading work with nearby ranchers for the use of their teams. On November 1 he went to town for groceries and on November 5 he washed and ironed. On December 5 he wrote, *"to cold to lay stone today worked for Jno Crawford,"* then he worked for Mr. Upham on December 7 and on December 11 mended his clothes.

November 29. I just put my name into a search engine and received an astonishing 33,700 results in .35 seconds. What would my parents think? My mother would have quoted the old saying about a woman's name should only be in the paper three times: when she's born, when she marries, and when she dies. My father would have snorted and said that the attention was probably all for my "worthless" writing.

Bill Snable wrote in his journal on December 18, *"Laid stone in forenoon and built cattle shed in afternoon for Mr. Upham. Christmas did not work."* What did he do? Was he invited to a neighboring home for dinner?

Through the years Bill "chored" for Mrs. Upham, worked on the city waterworks, butchered, worked on Mr. Upham's truck, shoveled coal, worked on his barn, *"halled"* corn, bought alfalfa seed from Jno Crawford, *"over-halled"* an auto for C. Arnold. On April 1, he figured he had worked 42 days for Mr. Upham and received a total of $41.65 in payment, about 99 cents a day.

His tally of his "bungalow" costs included $17.78 for lumber, $6.40 for windows, $25.30 for cement, 5 squares of roofing for $10. The grand total came to $103.33.

Exhausted by reading the cramped writing, I put the ledger aside and realize the sun is drawing long shadows before the grass. Time to walk the dogs.

November 30. I've been thinking about what I have learned in these months of reading and reflecting on the journals, letters, and other paper forming a family record. I hope I've observed in myself growth and change; I may be a kinder person today than I once was, and I've tried to learn from some of my less rational actions of the past. I am fortunate to live a comfortable life on this beautiful ranch and have gained enough perspective to choose more wisely than in the past how I spend my time.

And yet. And yet. I look down at the ranch yard and am still startled when I see the gap where an old shed stood until February, when Rick and his son tore it down and replaced it with a metal windbreak fence. The shed was there when I first moved to the ranch. I believe it was built by Jonathan Sharpe about 1912, the man who preceded us on this place, so it was over a hundred years old. Solid construction and our dry climate helped it last. The support posts were buried two feet in corral manure, but remained solid. Some wooden shingles had rotted. As I took photographs of their work, I heard Rick say, "I hate to waste stuff." Rick's crew gently took down the side walls, stabilized them with extra bracing, and built a pasture windbreak. They dug out the poles that held up the roof, all pitch posts worn smooth by the cattle rubbing on them, and stacked them to use again.

I know the shed roof would have collapsed if they hadn't taken it down, and these men didn't take the easy way of bulldozing it down and burying the remains or burning it. Respecting the work that preceded theirs, they gave time and energy to save what they could. I didn't let them see my tears, because I wholly approve of their actions. I must have been feeling sorry for change, for the fact that for sixty-odd years I have seen that shed in that place. Yet I cannot and must not resist useful change; that way lies Alzheimer's disease and crotchety old age, among other things.

— December —

December 1. Some days I'm just overwhelmed with creativity, but it's not specific to writing. I write good lines, scribbling in my journal; I put together a stew for supper and think over an essay outline. To someone watching, my movements might seem random as I race up the stairs to chop carrots, then back down to make a note, but it makes sense to me. Perhaps I work best in this kind of chaotic frenzy. Yet I'm always trying to make my office more orderly, to discard the unnecessary and file what I need to keep.

I've continued to puzzle over the Snable ledger, because on the second page of the journal, in brown ink, is this notation: "Pvt. W.E. Callahan, Bat. D 335th Art., Camp Pike, Ark." Eddie Callahan was my father's half-brother, born from his mother's first marriage. He ranched where my uncle Harold later lived, and he died somewhere on the ranch when a horse fell with him in 1942.

How did Eddie Callahan get Bill Snable's journal? And why did Bill Snable start his notes halfway through the ledger? I can tell Eddie took it with him to Camp Pike in 1917 and thriftily used it to take notes as he studied auto mechanics. Page after page provides basic information on the automobile engine, including detailed drawings of engine parts.

Somehow, the journal came back to South Dakota. I don't know where on the ranch Eddie died, but I've no doubt ridden past the place. I found the journal in my father's papers after his death, so no one is left alive who knows the story.

East of our house stands Bill's cabin, its roof beginning to sag. We've wrapped cable around it to try to hold the walls in place, but we can't afford to completely rehabilitate it. Vandals have stolen everything I remember in the house. A year ago a barn owl was nesting in the rafters above Bill's bedroom.

Eventually the cable will break or loosen, the cattle will get in through the windows and begin to rub on the door frames, and the walls will tumble, the rocks rolling out onto the prairie from which Bill gathered them. They will sink into the soil again, devoured by the land as was the man who gathered and stacked them, and the men who recorded important thoughts in this little journal.

December 2. Some days I lose track of the day, partly because of the sameness of winter days, and because in working with these journals, I am leaping from season to season. This morning I recalled and wrote about something that happened in June. I became completely immersed in the smells, the sounds, and the heat of the day. When I stepped outside to throw a ball for the dogs, I didn't put on a coat, and was momentarily stunned to realize it's 16 degrees with a sharp wind. Back inside, I flipped through my journal until I found a recent quote I copied about cold: "I'd never felt my bones as bones before—the dry clacking of the joints of the skeleton. My kneecaps, thin and brittle as sand-dollars, came to my attention first, followed by my wrists, knuckles, shoulder blades and ankles. I rattled as I walked. . . ."

We went to town in the afternoon for a couple of appointments but wished we hadn't. The driving lane was icy in shaded spots and curves, the passing lane snow-packed. Some folks were driving too fast for conditions, so we dared not go too slowly or some moron would cause an accident in passing. We have too much confidence in our big powerful cars.

December 3 The long high ridge to the south is gold in the weak sunlight, covered with deep grass. Throughout the grass run veins of white: snow-filled cattle trails where the cattle have worn their routes as they looked for sure footing. Nearly any trail walked by more than one cow once will show up under the thin snow we have now. As spring advances, many of these trails will be overgrown with green.

Staring at the hillside, I can almost see the bison that preceded our cattle here, placing their hooves between the same rocks, tracing the same seams as they find their way across the hillside eons ago.

I may have succeeded in putting into words some of what this country has meant to me in a new poem. I hope the title is not too obscure for readers; "autochthonous" usually refers to rocks and means "formed in place," or "originating where found." One source notes that such rocks "may have been intensely folded or faulted," a description I find appropriate to me.

AUTOCHTHONOUS

Sego lilies still grow
above the cedars hidden in the gully.
An old man in a rock shop
sells crystals across the road
from the field where the jet crashed
forty years ago; two pilots burned, screaming.
Sweet peas bloom in Black Gap.
Every spring we picked some
for my mother. When the land
was divided and house walls began to rise,
I said goodbye. But in that black shale
foundations shift and septic tanks slide.
Their owners learn more
about construction and sewage every day.
I may see those blooms each spring as long
as I live. A thunderstorm leaps the hills,
gallops toward me, rain riding toward
the Badlands one more time. The grass
that feeds those cows on the hill
twines through my flesh; the water
tastes of limestone
percolated through my bones.
This sun leathered my face;
this wind wove the wrinkles
at the corners of my eyes.
Each day that wind
erodes a little more.

December 4. The ash tree in my parents' back yard has been dying all my life, its black bark stripping away from the scar of a lightning strike so close to the house my mother said it nearly set her hair on fire.

Yet it lives, the naked trunk standing pale gray beneath black bark. Above my head, I can see buds that will swell with leaf as spring approaches. In Viking mythology, the ash tree was *Yggdrasil*, the Tree of Life. Healing wands and staffs were made from it, and it was brought inside to burn at Yule because it was believed to bring the light of the sun into the house on the darkest day of the year, the winter solstice.

Looking at the rough bark, I could see the scene. My mother often insisted they take a picnic to some cool spot in the Hills in summer. My father kept a spade and bucket in the car trunk and brought back dozens of pine trees, and undoubtedly this ash. He never tired of planting trees; if they died, he shrugged and planted more.

When we moved into the new house, probably in 1952, my mother wanted a city lawn. He dug dirt from a rich alluvial bank along Battle Creek and spread it over the front yard, then planted a weeping willow she could see from the picture window in the living room. By the time I was eleven or so, I could hide under the willow's sweeping branches while my mother called and called and called, "Linda! Linda! Come in here and do these dishes!" Eventually she figured out my secret place, and I started hiding in the wild willow bushes that grew below the well, east of the house. There I imagined myself an Indian or a fur trapper or a princess in her castle.

The ash still stands straight. I pat its trunk, thinking that my wrinkles are looking more and more like its creased bark, and turn to follow the dogs home from the morning walk. Passing the row of cedars my dad planted, I notice that each one has a rock beside the trunk, placed there to mark the spot when the tree was young, and to keep the cattle from stepping on the little trees if they got into the yard. Staring at a hunk of pink quartz as big as my head, I realize two things: that he saved especially beautiful rocks to mark his trees, and that he was the last person to touch that stone.

Everywhere I look are stones he hauled in from the pasture: edging the flower beds beside the house, squared into paths under the trees, and placed along the walls of most of the buildings in the ranch yard. He found fossils, petrified wood, quartz, and stones that might have been worked by someone who lived on this land a long time before we white people. If I knelt and patted any of those stones, I might almost feel the warmth his hand has just left behind.

I haven't cried for my father in years, but looking at the wall above my desk at the stern face he wore for the photographs he hated, I feel tears start in my eyes. I can't imagine that he could look into the future and see a time when I would be here, recalling his rock-gathering habit. On the other hand, he brought those rocks from places where they had lain since time began and placed them here with no expectation they would ever be disturbed. He knew someone would be here.

December 5. The sunset is covered by soft gray clouds so thick no pink or gold shows through. The white snow on the lake gradually fades into gray and the hills get darker and darker. As I watch, a white-gold globe rolls down the slope: a car on the highway, its headlights cutting a path.

In my father's journal I found this quotation: *"In the depth of winter I finally learned that there was in me an invincible summer."* How did my father, who studied engineering, encounter Camus? I need to find that summer.

December 6. I woke at three for the usual reason, and as I got back beside the bed, saw lights to the east and south, far away from where any lights should be in the middle of the night. After a moment I realized that I was seeing a train engine, its lights icy white against the thick black of the pasture dark. Slowly it moved out of the deep cut through the pastures where Jerry and I rode the four-wheelers, making no sound because the windows were closed. After a few moments, I heard the long hooting whistle where the track crossed a ranch road. As the silent train moved across the window, I slid between the sleeping dogs. Toby grumbled and Cosmo moved under my chin, and we all went back to sleep, warm and safe.

December 7. According to the radio's weather channel, the grassland fire index is high. The recent light snow has melted, and the grass is dry, so with the wind expected this afternoon, they are warning people not to burn garbage. Of course we never burn unless we have a solid snow cover, but I worry about people who toss cigarettes out car windows.

At sunset the light seems especially golden, lying across the long grass, and, because it sinks behind the hills before five in the afternoon, so much more precious than it is in summer.

I copied a fine quote into my journal today: "The butterfly counts not months but moments," said Tagore, "and has time enough." A good sentiment, but the butterfly does not write books nor cook three meals a day.

December 8. Not long ago, as Jerry and I walked up the driveway with the dogs, I picked up a black stone that was nearly heart-shaped. Jerry took it, and today gave it back to me polished so it truly is a heart. Words aren't his favorite mode of expression, but rocks will do.

Slipping the rock into my pocket, I thought of all the women I know, many of them my age, who pick up rocks, not "pretty" rocks but intriguing ones. I surround my plants with rocks I've collected, to prevent erosion and refresh my memories. The heart-shaped sandstone came from a Maine beach, the sand dollars and volcanic basalt from Manzanita.

Each rock triggers memories and expands my imagination. Sometimes they raise questions: Do I know anything about the material it's made of? What is its story? When I water the creeping thyme plant beside my bed, I sweep my fingers over it to release the scent, but I also smile at the deep blue fragments of glass I picked up in Maine, shining in the scallop shell from Manzanita; both memories take me away from the cold outside the window or the news I've read that day.

December 9. This morning the sun is shining, though the temperature is about 15 degrees. All around the western and northern horizon lies a thin, dirty-looking gray cloud. From long experience, I know this means a snowstorm is coming, perhaps even a real blizzard. I love that onomatopoetic word *"blizzzzzzzzard"* which makes the sound that tiny snow granules make as the wind skids them across the tops of frozen drifts. When we walk the dogs, the air is still and the sun feels warm. By mid-morning, the wind is beginning to pick up; the dry weeds are rattling, and a sheet of white so thin we can see blue sky through it has risen halfway up the sky. The white looks harmless, like a gauzy curtain, but it means snow, and possibly a lot of it.

December 10. Jerry brought in some branches he trimmed from the lower edges of the junipers and pine trees in the ranch yard, which makes mowing under them easier; I put them in a vase for our Christmas tree. We hung our oldest ornaments, recalling past Christmases. In the afternoon, he wrapped gifts for his family and for me. He wraps like an engineer: the paper is perfectly aligned, and his bows are artistic and beautiful. My wrapping is always haphazard, with crushed corners and re-used bows. He teases me about saving the paper, but I really have started throwing some away. My mother loved to recite the lineage of wrapping paper, saying, "I wrapped your birthday present in this when you were sixteen years old!"

At midnight, the moon is huge and golden; a great horned owl sits in one of the cedars near the house, calling, calling, calling. I can tell she's female because of the rhythm and tone of her song. I wonder if the light makes her hunting more difficult because her prey is more wary or if she's celebrating having had an ample supper of unwary rabbit.

December 11. When I first get up in the morning, around 4:30, I get out of bed without turning on a light and lean close to the south windows. If the night is overcast, I see nothing, but often I see that shimmering black ceiling with lights arranged in patterns whose names I don't know. No smog, no interruption, just lights coming from impossibly far away. Then I step out on the deck and look east, mostly to test the day, but also to absorb the fact that any light I see that direction will be a trespasser on land that is mine to care for.

December 12. I made green chile this morning, simmering it slowly until noon so the odor of that hot green fruit permeated the house. We'll eat a lot of it this noon, but have enough left so we can have our version of *huevos rancheros* tomorrow morning: fried eggs with green chile on top.

Christmas cards and laundry: I write the cards near where I am working on the photo albums, on a long table in the basement, so it's handy to jump up and check the dryer every now and then. I enjoy simply working somewhere besides the computer.

When I visited my mother during her last year or two, she was often sitting on the couch with a stack of books, reading. In the last few weeks of her life, when I tried to talk to her about the future, she often said, "Just leave me alone so I can read." She'd worked hard; she just wanted quiet time to read. Now I understand.

December 13. Wrapping Jerry's Christmas gifts today, I realized I haven't wrapped a gift for anyone else. No one. What does that say about me? I have friends, but we don't usually exchange gifts. Or perhaps our gifts to one another are spread out over the year that has passed, in the form of letters and telephone calls and books exchanged.

In a spirit of rebellion tonight, after immersing myself in my parents' journals so much this year, I ate several chocolate mint candies *after* brushing my teeth, courageously breaking one of the big parental rules after nearly seven decades and laughing at myself.

December 14. I've just had the sad experience of re-reading a cheery Christmas card I sent to a good friend with whom I've been corresponding for about fifteen years. Though our meeting was brief, at a conference where we were both speakers, we bonded instantly and immediately began exchanging holiday letters. We wrote only once a year, but our letters were individual, personal and long: about our writing, about our gardens and the things we were cooking. We exchanged recipes and recommendations for books to read.

Last year at this time, I was looking forward to receiving her holiday letter. She was a role model for me. In her eighties, she remained vigorous and curious, working on a long book that depended on her Greek translations. She'd given up a summer place to move full-time into her primary home, but she showed no signs of mental decline. Her most recent book had been about death and dying, based on her experience with her mother. It remains on my shelf, still unread.

In my Christmas card, I asked her how her new book was coming along and conveyed to her the compliments of a friend who was enjoying a previous book she'd written. I asked about an older book of hers I wanted to locate.

Why am I reading this message?

Because her card came back to me marked "UTF": Unable to Forward. My heart told me what that meant, but I spent considerable time online finding her address, mapping the location of her house and looking at a photograph of it, overgrown and neglected.

Then I found her obituary: she had died "after a long illness." But she had written to me the previous Christmas, filled with good cheer and encouragement. I cried for an hour.

I know this woman had a daughter; I know she maintained a hearty correspondence with many people and was beloved by many more who had read her books. I deeply sympathize with whatever hardships might have accompanied her illness and death. I can't help wishing her daughter had turned to my friend's, no doubt meticulously maintained, address book to let her friends know she was ill, or that she had died. Surely the death of a loved one requires this final effort: letting their friends know. Even a postcard would do. I think of the people who, like me, are enjoying this season and anticipating that annual letter. Surely letting them know would be the final gift to your dead loved one.

When my mother died, I found her address book to be fairly confusing, with scribbled-over addresses going back years. With help from my assistant Tam, who had spent so much time with her, I notified everyone whose address I could decipher. In return, I received letters about friendships that extended through fifty, sixty, seventy years, from people saddened by her loss, but grateful to hear from us, to know that, as one woman put it, "the song has ended."

December 15. At the basement entrance to our house stands our Iron Wall, a retaining barrier covered with old rusty tools.

The wall is a stack of railroad ties linked by rebar, holding the hillside back. We salvaged the railroad ties from the right of way cutting through our pasture. Free, sturdy, and impregnated with tar, they will resist water. They're also payback for the trash discarded by track crews, and the fires occasionally started by the trains.

On the wall we hang rusty antique tools discarded by my father in our personal pasture dump. I chuckle bringing these castoffs home; I can see my father shaking his head as I undo his work of throwing them away.

Already the functions of these tools are a puzzle to many visitors; I've used some and at least know the purpose of those older than I am. In another generation, they may all be mysterious. Whoever owns this house next might haul them to the recycling center or back to the pasture dump.

For now, I can look at worn horseshoes testifying to the miles walked by Bud and Beauty, the work team; the ornate legs of stoves and washing machines that supported the work of women for years. Here are the hobbles we used to keep the wild range cows from kicking while we milked them if they rejected their calves. Gears, hinges, pick heads, hand rakes, and the teeth from the dump rake I hauled for miles behind my tractor while my father stacked hay. Chains, pliers, chain boomers, winches that we used for everything from fastening gates to hauling pickups out of the mud. Swooping curves of metal with rings were probably part of a harness. Wheels with broken teeth hang beside stirrups pulled apart when a rider disagreed with an unruly horse.

On the basement bathroom wall hang objects I collected for their beauty as well as their utilitarian charm: hand-forged hooks for gates and cupboards in the barn; a hand-forged screwdriver, its length bumpy with hammer marks from that long-ago blacksmith. Jerry built a wooden rack to hold horseshoe and square nails, and hanging high are two vicious-looking hay hooks I kept handy when I lived here alone. Suspended in a row are several hand-forged nipple picks I used to clean black powder from the firing pins of my cap and ball weapons. Two calipers curve against the cream-colored wall. A small pitchfork was part of my child-sized equipment for feeding my horse, one of my first chores. A blackened leather bullwhip drapes over a nail next to a crude wrench used for wagon wheels. A horn weight was one of a pair we used to make a Hereford bull's horns grow into the gentle curve that was both attractive and safer for the cows and for us. Like the tools outside, many of these are ones I used.

Why have I brought all this stuff into to my home? Am I longing nostalgically for a return to the past?

Absolutely not. I admire much of the past, particularly as I lived it on this ranch, but not everything was better; my childhood was not idyllic. Watching the nationwide debate about vaccinating children against measles, I remember my parents' terrified hesitation about the polio vaccination—until a friend contracted the disease and was crippled.

By writing about this collection of rusty stuff, I have come to understand that I collect these things because they are another kind of journal, hand-forged, in a language that has been lost to many people today. They remind me to respect time, both in my writing and in my life outside of writing.

To make hand-forged gate hooks for our barn, for example, required that the blacksmith own and know how to operate a forge, collect the appropriate tools and supplies of metal. One didn't fire up a forge to make one hook, so he probably had a number of forging jobs to do at once. He may have taught himself to work iron to save money on his own ranch. Or perhaps someone in this neighborhood was a blacksmith, and the others traded beef for what they needed.

The blacksmith had to select the steel, then build a small fire in his forge to ignite the coal. He used a bellows or a hand-cranked blower to add air to the fire until the coal was burning well. Then he thrust the steel into the coal to heat, moving the bellows to add air.

My partner Jerry has worked at becoming more skilled as a blacksmith since he retired. He tells me that the blacksmith must heat the steel just enough. "Reddish-orange to bright orange is good," he says. If the steel becomes white hot, it will simply burn up.

When the steel is ready, the blacksmith removes it from the fire and begins to shape it. Several of these hooks have a twist in the middle, which might add a little lateral strength, Jerry says, but likely was pure decoration. So the hook was created by a man who had plenty to do, but took time to add a flourish to his work.

Besides materials and skill, a blacksmith needs time and patience. When Jerry spends a day blacksmithing, he's already thought about the project while doing other work, eating good food, watching movies, and playing Scrabble. He collects the materials and waits for a day when the woodstove in the blacksmith shop will heat the place enough so he can work without freezing solid. Then he heats and reheats and hammers until he has created what he visualized. I don't know what he thinks about, but I know he thinks because hammering iron requires patience and allows time to consider other matters.

The metaphor between writing and blacksmithing is obvious: we both spend our days planning and have the ability to imagine the finished product from unfinished materials. I heat an idea in my mind's forge until I can beat it into shape; watch it cool as I revise, then pound and sweat and mumble until I'm satisfied. Sometimes the idea burns away as I overwork it.

Like the books in my study, these hand-shaped tools symbolize the patience required by any job done well.

December 16. Another Star Wars movie has been released. After I told Jerry I didn't especially want to go, he said, "Do you remember how much George loved them? You said when you went, you were practically black and blue when you came out because he kept elbowing you at the good parts?"

I was stunned. How could I have forgotten something so important about the husband I loved so much? Jerry remembered George had told him that the movies were exactly the way he believed space exploration would be. After George died, I dreamed I found his clothes on the entrance road with a note saying that he was from a long-lived, star-traveling race and had to visit other planets; that when he came back, I'd be long dead. The dream no doubt arose from my memory of his Star Wars fantasies, and yet I had forgotten. I feel disloyal to George.

Still, Jerry and I are committed to one another, and part of our responsibility is embodied in living for today. We've each known couples destroy good relationships by comparing them to previous spouses or lovers. When I shut the door on my life with George, I didn't empty the room that was our life. Everything is there, in my brain or heart or a trunk or two, but I believe it's better not to get those memories out and swoon over them. In the same way, I've kept my parents' journals and my own; the writings preserve memories, but they don't need to be aired often.

December 17. Walking the dogs this morning, we saw a large set of canine tracks in a circle, the snow sprayed out as the animal turned, clearly chasing a rabbit not ten feet from our sidewalk. Looking closer, we realized the tracks are much larger than the coyote tracks we've been seeing. This critter left tracks more than four inches in length, not necessarily large enough to be a wolf, but certainly bigger than most of the coyote tracks we've seen. This is the kind of event that makes me question my devotion to keeping the native wildlife healthy: if this is a wolf, it's way too close to our house for comfort. Even a large coyote chasing rabbits up the front walk is unnerving. When I let the dogs out in the morning, I usually step outside with them and look around, then hasten back inside. Very easily a coyote could leap into their pen and have them before I could get back to the door. One of us goes with them when we let them out just before dark, and then just before bed, Jerry goes out with them, sometimes smoking a cigar in the back yard.

Reading one of my favorite mystery authors, I find a paragraph that seems to apply to what I've been thinking and writing.

The past, suggests Jane Langton, is weird and unknowable. How can we presume to know anything about it when "every gesture of a hand, every habit of talk, every simple unconscious action was shrouded by the curtains of the years that lay between?" Each year that passes makes the memory fade a little more.

Yet we spend a great deal of time trying to understand the past, both officially, in history books and long analytical writings, and unofficially by looking at photo albums and poring over diaries. Surely I have spent more time than most at that activity this year. What have I learned?

I've remembered many details about my father and mother, not just details like where she went to school or what brand of vehicles he preferred, but gestures and expressions that I couldn't describe in words, the idiosyncrasies that defined them as individuals. No one else remembers those particulars and when I am gone, my parents will be more completely gone as well.

My mother's brothers' families and I have photos of my mother as a teenager and young woman. I have boxes of her letters and journals, as well as photographs she took during her lifetime, though many of those are of her cats. These documents could be used to create a well-rounded picture of the life of a woman of her era.

I have almost no such documentation of my father; he has almost vanished from the world already. His siblings are all dead. I have one family photograph in which he appears and in which his withered arm is evident. I wonder if he destroyed other photos, perhaps when he acquired them after his mother died. I can't even get confirmation from the University of Nebraska that he graduated from that institution with a degree. In my office are two photos of him, both unsmiling, because, as I am the only one left to remember, he hated having his picture taken. Still, he looks straight out of those frames, like "the man in the glass" in the poem he recalled, facing whatever his life brought him.

Recently when I left a casserole in the oven a little too long and burned it on the bottom, Jerry said, "But that's just the way I like it!" I laughed until I nearly cried, because my father used to say that when my mother overcooked a meal and Jerry remembered.

December 18. A friend from eastern South Dakota recently wrote to remind me that I introduced her to the kestrel by identifying a bird she had seen for some years on the wires along their road. She writes of first seeing the kestrel sitting in a ginkgo tree in her garden, unafraid.

> *"Next morning it scattered the birds at the nearby feeder and soon caught one. I was quick to watch with the binoculars, impressed and delighted it perched within perfect view. Of course it returned the next day, and the next, and now I found myself unsure of my regard. I did not like that I could be so cavalier about whether or not it*

caught and ate from the multitude of English sparrows but was anxious the lone male cardinal might be devoured. . . . Well, the birds have adjusted and when none are about I know the kestrel is, and when they are feeding I know he is gone but not for long. Also the dynamics at the feeders has changed. The birds feed differently, never lingering at the feeder or beneath it, darting into the nearby shrubbery between seeds."

Just as my observation of the kestrel made her more aware of her habitat, her observations have made me notice that a kestrel which hunts the windbreak near our house has trained our sparrows too; when we walk into the windbreak, we hear twittering but rarely see the birds, safely gobbling juniper berries under the trees.

Predators train their prey to be more alert. Yet humans, comfortable at the top of the food chain, usually forget such dangers until they walk in a park where buffalo and mountain lions roam. Today was so warm and sunny that in the afternoon, with the thermometer at 50 degrees, we drove up into Custer State Park and hiked a couple of miles on a winding, dipping forest trail in sunlight and shadow. Alertly, we watched behind and scanned ahead as we strolled among the buffalo dung and mountain lion paw prints on the trails, seeking evidence of their predatory presence in the journal of the earth.

December 19. In a local grocery store Saturday, a scruffy-looking gentleman who was finishing shopping offered me his cart.

He supported himself with a gorgeous twisted wood walking stick as I took the cart. "Beautiful walking stick!" I said, "Did you make that?"

"No. The Lord did."

With a smile gentle as a blessing, he went out the door.

Beautiful day. Black Hills Raptor Center has let me know that the hawk I turned over to Game, Fish and Parks is an adult Cooper's hawk, the first I've ever knowingly seen. They don't yet know what killed it; that will have to wait for the taxidermist. These accipiters usually nest in deep woods like the Black Hills or cottonwood groves in the Badlands. They are bird-hunters, and can reportedly outfly any bird through the trees, ducking and side-slipping through breaks where no other bird could follow. BHRC directed me to a YouTube video that made me dizzy, a Cooper's hawk wearing a camera. Apparently Air Force pilots have said that they have no plane capable of the kind of maneuvers this bird executes in its search for prey. I am so glad to help educate people about these wonderful birds.

December 20. Coyote howls spiral into silken blue dark. The moon's a silver enigma.

Is it simply aging, or weariness, that makes me happy to serve in the kitchen when the local history group holds a fundraising dinner? Am I simply being lazy when I decline an invitation to speak a thousand miles away in return for travel expenses and a very small honorarium?

I believe that I'm being true to my own need to write and think in uninterrupted peace in my own home, which is stocked with materials to aid my writing and thinking. I've spent years traveling all over the country at the behest of various groups to whom the work I presented was perhaps only part of one day's entertainment. Respect for my own work requires me to stick to it. What I write may not last beyond my lifetime, but written work often has a longer life than the spoken word.

December 21. Winter Solstice will occur at 4:03 P.M. We are going into the Hills to see friends, visit beside their bonfire to appreciate the change in seasons and eat soup and bread: a perfect way to celebrate the shortest day and longest night, this enveloping darkness before winter. The short day leading to the year's longest night promises that spring will come.

I've read that some church congregations have taken to celebrating the solstice, recognizing that their beliefs are connected to earlier earth-based religious practices. In one instance, the first half of the church service is marked by the lights in the nave growing progressively dimmer while the second half features the lights growing brighter. In the middle, however, is twelve minutes of complete darkness and silence. Sitting quietly in the dark for any length of time, says one pastor, is a radical act in our society.

Usually I welcome the solstice by wrapping up warmly and sitting in one of our chairs on the deck, looking up at the stars visible as long as I keep my back to the north and west, where the city and the subdivisions announce their presence with glaring yellow lights. Facing east and south, I lean back and look up at the Milky Way and the other stars scattered over the black above me. I think about the year that is passing away, concentrating on what has been good about it. Sometimes I ask myself how I can improve the year to come. Sometimes I simply sit staring out at the dark, listening for coyotes' howls in the wind's voice, or perhaps the cough and rumble of some buffalo who roamed here a hundred years ago.

The important thing, I remind myself as often as I can, is to make everything I do part of the good day that I want to enjoy. Each step, vacuuming, cooking, working on a poem, folding laundry, is part of a day that is good because I am alive and able to do it. So today, celebrating this shift into darkness that will lead to spring, I have written cards to friends, cooked a dinner of cod, potatoes and peas, and cleaned house. Having done all those things, in the spirit of solstice, I sat down to read for an hour, and to write in my journal, as the afternoon slid away over the Hills.

December 22. Blustery chinook wind blowing today, though the temperatures so far are only in the 40s. Walking under the cedar trees north of the retreat house, Cosmo came upon the tracks of the big coyote, or the coy-dog hybrid, or whatever it is, and while he was studying those intently, an owl dropped down over our heads and flew up the draw. I heard just one *swoosh* of the great wings, and that only because the owl was just a few feet away when it got into open air.

What would my parents think of their house being a writing retreat? My father, in his later years when deterioration had turned him mean, wouldn't like it, but his shade, if it haunts his ranch, is probably in the pasture checking grass. I think my mother would be delighted, though she'd sneer at the uncarpeted floors. Several writers have said they felt a benign spirit in my parents' bedroom, and I know she'd love the books. Now I'm going to see her there: stretched out on the couch where she spent her happiest days, with pillows and books on her lap.

December 23. Frost covered the interwoven spider webs under the deck this morning, and in the afternoon they began falling with its weight: long spikes of frosty web.

On our morning walk, we were startled by the flight of half a dozen crows that flew to the trees. A murder of crows! They cawed and circled and flapped and squawked, but I never could see what agitated them. Perhaps a visiting cat, hunting.

The retreat that is ending today has been a success, I believe. I've talked a lot and provided the writer with a stack of handouts four inches tall. She's absorbed everything like a sponge, but without losing her own viewpoint and perspective; we've had some vigorous discussions about how much she wants to tell in her memoir about her very difficult life. I find it hugely satisfying to know that I'm sending her home with as much written material as she would have gotten in a semester-long class; so she has all that she needs to keep writing—except the persistence and motivation which I hope I inspired in her.

December 24. I made a fruitcake to eat for breakfast tomorrow morning, one of my favorite very dark, very rich ones with no candied fruit, but dried dates, figs, and cherries.

Working on my album of photographs for the year, I got to May before the electricity went off at 2 P.M. I was digging out the candles and lanterns when it came back on at 2:20. The photo album work I do at Christmas always makes me think of the past and future. Tonight we'll get out albums from several years and look at them, smiling at memories. I can't help thinking: Where will I be next year? In five years? Will I still be alive in ten years, able to look at these records? Will I remember, or will I be lost in a world that seems strange and even hostile? I hope that my body fails before my mind, but some of my memory lapses lately worry me.

I usually try to present the album to Jerry for both of us to look at on Christmas morning, but I am a long ways from finished. Perhaps for New Year's Eve.

In the evening we went to my cousin's house in town for chili and oyster stew, a Christmas Eve tradition now. We won't know anyone there except them, but it's a quiet way to spend the evening and makes us both remember other Christmases with more or different family members.

December 25. At 6:55 this morning, the full moon sank below the Black Hills just a few minutes before sunrise began to turn the eastern horizon pink. This is the last full moon to appear on Christmas until 2034. If I am alive, I will be in my nineties. Will I watch that moon rise from somewhere else? If so, who will watch the moon rise from this ranch?

Around nine that morning, I poured myself a wee dram of Scotch, with a glass of bourbon for Jerry, which we sipped with fruitcake and coffee while we slowly opened our gifts, including stuffed toys for the dogs. Then we packed away the paper and ribbons we could save and piled our gifts back under the tree to admire. We each made calls thanking friends and relatives for thinking of us and listened in a little on other Christmas celebrations.

All morning, we could smell the ham slowly roasting in the oven. At midmorning I peeled potatoes to tuck around it, and dipped out a little juice to put with the green beans from our garden I'll sauté lightly just at noon. I broke up some lettuce for each of us and made some of the blue cheese dressing Jerry likes.

In the afternoon I sat in the sun below the deck, sheltered on the north and west by walls, warm in a black flannel dress even though it was 40 degrees. My good friend Grete died several days ago after a long and cancerous illness, and I've been unable to concentrate on my writing very well since then. I was invited to help write her eulogy but refused. I can't put the reason into words, but I have been unable to articulate my sense of loss for several days. So I sat in the weak winter sun, soaking up heat, closed my eyes and thought about the time we traveled to the southwest together, how often we laughed. Once she screamed, "Stop! Stop the car!" and when I did she ran to the borrow ditch beside the two-lane highway. I followed in time to see the skeleton of a dead elk. She was already dragging the huge antlers toward my car. Of course I helped her, both of us panting with haste and the weight of the bones, me gasping, "I'm sure this is illegal."

"So hurry up," she said.

We covered the antlers with a blanket and hauled them to her house. After a few years, she brought them to me, saying they belonged in the open. So now they are inside the rows of my windbreak juniper trees, slowly desiccating. I'll think of her every time I walk by them, not of her death, but of her laughing and refusing to listen to my protests. Her eulogy.

Chocolate and mints to end the day.

December 26. Since the dump was open, we loaded up all our garbage and recyclables and made the trip, our practical celebration of the year's end. Another unwritten journal?

We had leftover roast beef in tortillas for lunch, ham sandwiches for supper. We love ham and beef roasts because they keep on nourishing us for days in leftovers. Of course I have to say that good writing is the same: just keeps on providing for the mind as leftovers do for the body.

I also made Jerry's traditional birthday cake: flourless chocolate, which cuts the carbohydrates but allows him to enjoy the rich, silky, dark flavor of chocolate. I love it too.

While he took a hot bath, I got his birthday gifts out of all the nooks and crannies where I'd hidden them and arranged them on the coffee table.

December 27. Jerry enjoyed his birthday working in his shop—building my new writing desk—and I enjoyed writing.

I roasted a turkey with dressing for his birthday dinner, with all the trimmings he enjoys—white potatoes (not yams!), gravy, homemade bread, and our favorite jellied cranberry sauce.

December 28. When I come in from walking the dogs, I sit in a warm room to work, but the cold seems to be inside my ankles and thighs. I can picture those long bones as blue-white ice, deep inside my pink flesh, chilling me from the inside out. Only a hot bath warms me up enough to stop shivering.

I un-decorated the tree branches today, putting away the ornaments, some of them from my childhood. As I looked at Christmas cards, I updated my list, with fewer people each year.

For salad this noon, I cut up apples and cheese; Jerry prefers the tart green Granny Smiths and I like something sweeter, like Jazz Jonathans. I take a bite of cheese, then bite of apple, just as I remember my mother doing on picnics in the park in Rapid City. The journal of memory.

Recently I read that medical research shows it's possible for some information to be inherited biologically through chemical changes that occur in DNA. I have worked to overcome my fear of spiders and love to eat cheese and apples together; my mother loved cheese and apples together and was terrified of spiders; I have inherited both traits. How many things that we do are products of events experienced so early the memories exist not in our minds but in our blood? Is some of our behavior evidence of the unseen journals in the folds and shadows of our frontal lobes?

December 29. At 4 A.M. the temperature was 12 degrees; at 8 A.M. it's 3 degrees with a vicious wind. The dogs couldn't walk long because their feet collected ice, and they looked at me, holding up frozen paws. We could hear the grouse chuckling in the juniper trees.

December 30. Minus 10 at 4 A.M. Viciously cold, with a prediction of wind chills of 40 degrees below zero tonight. Our propane is getting low, so we hope the truck can get through tomorrow. With snow on the ground, Jerry was able to burn garbage. For supper: leftover Christmas ham in bean soup.

Today I contacted a nearby college library with a statewide historical collection that has happily agreed to accept my donation of the journals, photos, and documents I've collected about my parents, the ones I've been reading all this year. This seems a fitting end to my struggle with the hidden meanings; perhaps some future researcher will decipher my mother's journals and find a significance in them I've been unable to see. Along with the journals, I'll deliver the collage I made of photographs of Mother; I spent today identifying and dating the photos.

I've also signed and mailed a deed of gift for my own collection to my alma mater, the University of South Dakota, donating collections of publications and reviews of my books, boxes of unpublished writing, photographs and letters, copies of my talks, and finally, my journals, which will not be delivered until after my death. Already, though boxes of all these things are still in the basements of both houses, I feel as if I've shifted a burden. No more poring over the past; I'll try to concentrate on the future, and on new writing.

December 31. I've made no firm decision about the future of the ranch, but I told Rick today that I'm likely to sell him the rest of it, though possibly not until age or ill health forces me to leave. I'm keeping my options as open as I can. I'm unlikely to buy a horse and saddle and a few cows, but only the dead have run out of choices.

I sold my east pastures about a month ago to Rick, who bought Harold's ranch. He and his son seem to manage the grasslands well; the grass looks as good, or better than it did when my father was in charge. If I sell them the rest of the ranch, that choice will be wise, or at least as good as I can do, given the knowledge that I have. Rick earlier bought Harold's ranch, and I find it somehow fitting that the Hasselstrom Brothers ranch will be joined again—the land that Harold and John divided when my father married my mother because she couldn't get along with Harold's wife Josephine.

For lunch I made fried rice with turkey and a handful of shrimp. We'd been discussing friends who were going out tonight for "surf and turf"—lobster and steak—so Jerry opined that our lunch was "surf and perch."

Today we can see the highway for the first time in three days of blowing snow. Interstate 90 is closed, which will mean hundreds of people intent on getting home after the holidays, or to some New Year celebration, will be out in their four-wheel drive vehicles. When they get stuck or drive into the ditch or roll their cars, they will use their cell phones to call for help. Hundreds of emergency workers who are trained for these conditions will risk their lives and spend thousands of public dollars rescuing people who should have had the elemental intelligence to stay home.

In celebration of the end of the year I spent a lot of the day cleaning out files, dumping copies of letters that are no longer relevant, or turning them over to be recycled by printing on the other side. I finished our personal photo album, so after a dinner of spicy black bean soup and pumpkin pie, Jerry and I sat on the couch and reviewed our year. A waning gibbous moon hung on the horizon, dull gold.

This afternoon a couple from Nebraska arrived to begin their New Year's retreat, a time of rededication to their marriage, as well as working with me on their two separate books. What a fine way to end one year and commence another.

At different times, I've thought that this book should end with dramatic pronouncements and portentous thoughts, but it seems more true to let the sun set slowly on it, as Jerry and I pursue our ordinary quiet lives, enjoying peace and contentment. We will go on living day by day by day, enjoying the details of the lives we lead here. "One day, I know, it will be otherwise," wrote Jane Kenyon. We'll deal somehow with whatever occurs, leaving some evidence on this landscape we inhabit, to be seen and appreciated—or not—by those who come after us. The journals of paper will be tucked away in archives, but the journal of blood and grass remains.

Acknowledgments

Thanks to Molly Bredehoft, Laura McCormick, and Patsy Parkin for taking exceptional care in their editorial work and proofreading for High Plains Press. I've never met you, but I've learned a lot from our conversations in the margins of this book.

Thanks to Tamara Rogers, without whose hard work and good cheer, I could not do my own work while conducting retreats and offering online consultations to other writers.

To Jerry for more than I can express.

Thanks to Nancy Curtis, who manages to be both a tough editor and an enduring friend, and who helps Jerry and Tamara keep me from making a complete fool of myself.

Bibliographic Essay

January. I first saw the quotation by Isak Dinesen mentioned on January 1 in *The Human Condition* (1958) by Hannah Arendt (University of Chicago Press, 1958). John Steinbeck wrote of the connection between story and "hearer" in *East of Eden* (Penguin, 1992). I quoted Tim Sandlin's *Jimi Hendrix Turns Eighty* (Oothon Press, 2014) on January 14 and again on February 8. In the entry for January 15 appear quotes from two of my favorite writers on rural living, Susan Wittig Albert (*A Wilder Rose,* Persevero Press, 2013; Lake Union Publishing, 2015) and Verlyn Klinkenborg (*The Rural Life,* Little, Brown and Co., 2002).

February. On February 9, I quoted from Ann Haymond Zwinger's *Shaped by Wind and Water: Reflections of a Naturalist (The World as Home)* (Milkweed Editions, 2000). On February 15, I allude to a talk to the Hermosa Arts and History Association by Dan O'Brien, *Buffalo for the Broken Heart: Restoring Life to a Black Hills Ranch, Wild Idea: Buffalo and Family in a Difficult Land,* and other books. Barbara Kingsolver's *Animal, Vegetable, Miracle* (HarperCollins, 2007), mentioned on February 16, is educational but I still wish she suggested teaching school children how cattle are raised in the grasslands. You can find more information about ground truthing, first mentioned February 20, at www.groundtruth.in (the Ground Truth Initiative). Rainer Maria Rilke's comment on love quoted on February 23 appears in *Letters to a Young Poet.* (Angelico Press, 2014).

March. Kahlil Gibran's *The Prophet,* (Alfred A. Knopf, 1923), quoted on March 1, was published when my mother was an impressionable fourteen years of age. Jane Kenyon's poem "Otherwise" appears in *Otherwise: New and Selected Poems,* (Graywolf, 1996). On March 2, I refer to Peter, Paul, and Mary's "The House Song" from their *Album 1700* (1967). The story I mentioned on March 3 was "Chile Earthquake Altered Earth Axis, Shortened Day," from News.nationalgeographic.com (2010). I was thinking about the Sixties again on March 7, humming "Carry That Weight" from the Beatles (*Abbey Road*, Apple Records, 1969). Gail Sheehy's book *Passages: Predictable Crises of Adult Life* (Ballantine, 2006), mentioned on March 15, helped me make a crucial decision. My William Faulkner reference on March 23 is to his trilogy, *The Hamlet, The Town* and *The Mansion*, (Random House, 1940, 1957, 1959). The essay I was writing on March 25 was finally published as "Beulah Land" in *A Road of Her Own*, Marlene Blessing, editor, (Fulcrum, Inc., 2002). On March 29, I quote Jane Siberry's "Bound by the Beauty"; find it in "Gift: A Jane Siberry Sampler" (Wing-It Music, 2010).

April. The "gigantic local history" book (920 pages) cited on April 1 is *Our Yesterdays* (Eastern Custer County Historical Society, 1967-1970). On April 6, I recall meeting Meridel Le Sueur (1900-1996), an American writer of the proletarian movement in the 1930s and 1940s, and an activist for workers' and women's rights. She made this statement at a conference at the University of South Dakota in Vermillion, but I'm not sure of the date. The Moving Wall is a half-size replica of the Washington, D.C. Vietnam Veterans' Memorial which has been traveling around the country since 1984.

On April 13, I refer to Abigail Tucker's article, "Voles in Love," (*Smithsonian,* February 2014). On April 19 I mention Meghan Stump's "Web of Intrigue: New Spider Species Spotted on Fort Pierre National Grassland," (blogs.usda.gov, April 15, 2014). My April 24 pun, "Two Dog Night," refers to the name of the American rock band, Three Dog Night, formed in 1967 and named for the practice of Australian indigenous peoples of sleeping with one, two, or three dogs, depending on how cold the night is.

May. On May 3, I quote part of a poem by Grace Noll (later Crowell), an American who wrote more than 5000 poems and 30 books of inspirational verse. Edward FitzGerald, self-styled author of *The Rubaiyat of Omar Khayyam,* supposedly translated more than a thousand poems by Omar Khayyam from Persian. He called his versions "transmogrifications," and admitted he wrote many of the verses. Elizabeth Barrett Browning's collection of 44 poems, *Sonnets From the Portuguese* (1850), was written for her husband, who published them after her death; she believed they were too personal, so the title suggests translations rather than original work. I mention Bryan's speech, recorded 25 years after it was presented, on May 6, along with the *Rapid City Journal* reprint of Christopher Doering's "Population shift weakens rural America," originally published in the *Sioux Falls Argus Leader* (January 14, 2013).

The song in my head on May 12 was Joni Mitchell's "Woodstock," which some say was written after she heard her boyfriend Graham Nash's narration of his trip with Crosby, Stills, and Nash. On May 13 and 14 and again on September 27 and 28, I refer to Wendell Berry, The Way of Ignorance and Other Essays, (Counterpoint, 2006), one of his many excellent reflections on rural life. To learn more about fireflies like those I mention on May 16, see www.firefly.com. The May 30 comment commonly attributed to Charlie Chaplin is quoted in his obituary in The Guardian (December 28, 1977).

July. Some of the information on birds mentioned on July 9 came from www.allaboutbirds.com. Clearly the poem "Horse Sense," which I quote on July 14, is old since my father memorized it; I found a second verse at www.poetrynook.com but no indication of its original appearance. To learn more about Prairie Rose Henderson, mentioned on July 21, see www.naglewarrenmansion.com. The two essays I cite on July 25 concerned a blizzard, published May 19, 2011, and an antelope, published January 6, 2012 in Orion magazine.

August. On August 1, I consulted *The Sibley Guide to Bird Life & Behavior,* David Allen Sibley. (Alfred A. Knopf, 2000). On August 21 I mention Joseph Bottum's comments about South Dakota temperatures from his *The Christmas Plains* (Image Books, 2012).

September. On September 1, I quoted Verlyn Klinkenborg's *The Rural Life* (Little, Brown and Co., 2002); my copy is filled with underlining and sticky notes because I find so much to admire. "My favorite botanist," mentioned on September 7, is Cindy Reed, president of the Great Plains Native Plant Society, established on my ranch. Information about threats to the monarch butterfly mentioned September 7, appeared in "Monsanto May Finally Kill Off the Endangered Monarch Butterfly" (www.livinggreenmagazine.com,

March 3, 2014). "Broken Glass," which appears on September 9, was published in the annual anthology of Story Circle Network, *True Words* (2013). *No Place Like Home,* noted on September 9 was published in 2009 by University of Nevada Press.

After writing on September 19 about "The Guy in the Glass (The Man in the Mirror)," I learned that author Dale Wimbrow published it in 1934, when my father was in his twenties. "Strike the tent…" is from "The Bison Track" by Bayard Taylor. "The moving finger…." is another quotation from Gibran's *The Prophet.* On September 21 I mentioned Sayer Ji's article, "'Killer Germs' Obliterated by Medicinal Smoke (Smudging), Study Reveals" (www.green medinfo.com, May 31, 2015). The September 22 reference is to Annie Dillard's *The Writing Life.* On September 23, I refer to information from J. Wayne Burkhardt in RANGE magazine (Spring 2016) on the evolution of grass. The poem "Clara: In the Post Office," can be found in my book *Roadkill* (Spoon River Poetry Press, 1987). On September 30, I recalled Mother singing Kenny Rogers' hit "The Gambler" from the album of the same name (United Artists, 1978).

October. On October 3, I woke up singing John Fogerty's "Rock & Roll Girls," (*Centerfield,* Warner Brothers, 1985) but by October 5 I was humming along with Jackson Browne's "Running on Empty" from the album of the same name (Asylum Records, 1977). Searching online for information after my October 7 memory of crying when I heard Rod Stewart's "Forever Young" in 1988, I discovered that when Stewart realized that the song was similar to a Bob Dylan song with the same title from 1974, he consulted with Dylan; the two agreed to participate in the ownership of the song and share Stewart's royalties. On October 9, I quote Dylan Thomas's poem, "Do not go gentle into that good night." (*The Poems of Dylan Thomas,* New Directions, 1952).

On October 10, my brain was setting my life to the tune of Bob Seger's, "Against the Wind" (Capitol, 1980). Jim Kelly's book *The Moon Tunnel* (Minotaur, 2005) provided a line I quoted on October 11, but by October 12 my brain was repeating lyrics from Jackson Browne's "Take It Easy," the Eagles' first hit single (Asylum, 1972). On October 16. I recalled memories from both my parents about "Red Wing," written in 1907 with lyrics by Thurland Chattaway and music by Kerry Mills. I loved reading Robert Mac-

Farlane's *The Old Ways* (Penguin Books, 2013), mentioned on October 24, though his reflections don't apply to paths in my neighborhood. Nor do most of Henry David Thoreau's observations, though *Walden*, from which the October 26 quote is taken, inspired me in both writing and walking. The hotel we visited for Halloween may have been the Glendale, built by Ed Stenger in 1886 as one of the first buildings in Hermosa and torn down in 1968; a photograph of it appears on page 620 of *Our Yesterdays.*

November. My reference to St. Augustine's words on November 3 was possible because my memory had been refreshed by Greta Austin's article, "St. Augustine and the Hall of Memory" (*The American Scholar*, Winter 2012). I'm indebted to Eric A. Powell whose article "Letter From Montana: The Buffalo Chasers," (*Archaeology*, October 14, 2014) demonstrated to me on November 7 that we probably have primitive bison drive lines on Badger Ridge south of our house. Referring to "one of my favorite mystery writers" on November 25, I'm speaking of Sue Grafton, from whose *W is for Wasted* (B.P. Putnam's Sons, 2014) I quoted.

December. The note about cold I copied into my journal on December 2 came from Jonathan Raban's *Bad Land: An American Romance* (Pantheon Books, 1996). My weighty *American Heritage Dictionary* (Houghton Mifflin, 2000) a Christmas gift from Jerry the year it was published, supplied the definition of "autochthonous" that inspired the poem on December 3. The poem was published in *Dirt Songs: A Plains Duet* (The Backwaters Press, 2011), fifty poems each from me and from Twyla M. Hansen, Nebraska poet laureate. My December 5 quotation from Camus, found in my father's journal, still puzzles me. The source is Albert Camus' "Retour to Tipasa 27," a lyrical essay contained in the 1954 book *Summer (L'Été).* My father did not read French and he did not use a computer, so he neither read the book nor found the quote on the Internet; he did, however, read many news magazines.

Nobel Laureate Rabindranath Tagore (Indian poet, 1961-1941) is the author of the butterfly quote I copied on December 7; it is from a poem titled "I Touch God in My Song" in the book *The Fugitive* first printed in 1918 (Cosimo Classics, 2005). Jane Langton's *Steeplechase* (St. Martin's Press, 2005) supplied the quote about the past I copied on December 17. My friend C. B. (Vermillion, South Dakota) made the comment on the kestrel included for December

18. On December 28, I quote Linda Geddes' article "Fear of a smell can be passed down several generations" (*New Scientist*, December 1, 2013). Jane Kenyon's poem quoted on December 31, I first mentioned on March 1.

Archived sources. The journals of Mildred and John Hasselstrom, Harold Hasselstrom and Bill Snable, the letters of Cora Belle (Pearcey Baker) Hey, Charles Hasselstrom's ledgers, Jonathan Sharpe's papers, and many of the other documents referred to in this journal are now in the South Dakota historical collection at the Devereaux Library, South Dakota School of Mines & Technology, Rapid City, S.D.

My own papers are available in several collections, which will be augmented after my death. For more information, ask about the Hasselstrom collections at the following institutions:

Custer County Historical Society, Custer, S.D. 57730;
Archives and Special Collections, University Libraries,
University of South Dakota, Vermillion, SD, 57069;
American Heritage Center, University of Wyoming,
Laramie WY 82071-3924;
The Hermosa Arts and History Association,
www.hermosahistory.com

Nature Centers. The Great Plains Native Plant Society (www.gpnps.org) and The Black Hills Raptor Center (www.blackhillsraptorcenter.org) rely on donors to continue their work of educating the public about native plants and raptors.

Further Reading

Brown, Lauren. 1985. *Grasslands: A Comprehensive Field Guide.* New York: Alfred A. Knopf.

Cassells, E. Steve. 1986. *Prehistoric Hunters of the Black Hills.* Boulder, CO: Johnson Books.

Densmore, Frances. 1986. *How Indians Use Wild Plants for Food, Medicine & Crafts.* New York: Dover Publications.

Gilmore, Melvin R. 1977. *Uses of Plants by the Indians of the Missouri River Region.* Lincoln: University of Nebraska Press.

Halfpenny, James, and Elisabeth Biesiot. 1986. *A Field Guide to Mammal Tracking in North America.* Boulder, CO: Johnson Books.

Johnson, James R., and Gary E. Larson. 1999. *Grassland Plants of South Dakota and the Northern Great Plains.* Brookings, S.D.: South Dakota State University College of Agriculture and Biological Sciences.

Jones, J. Knox, Jr., David M. Armstrong, and Jerry R. Choate. 1985. *Guide to Mammals of the Plains States.* Lincoln: University of Nebraska Press.

Kindscher, Kelly. 1987. *Edible Wild Plants of the Prairie: An Ethnobotanical Guide.* Lawrence: University Press of Kansas.

———. 1992. *Medicinal Wild Plants of the Prairie: An Ethnobotanical Guide.* Lawrence: University Press of Kansas.

Robinson, Jo. 2004. *Pasture Perfect: The Far-Reaching Benefits of Choosing Meat, Eggs and Dairy Products from Grass-Fed Animals.* Vashon Island Press, Washington.

Royer, France, and Richard Dickinson. 1999. *Weeds of the Northern U.S. and Canada.* Edmonton, Alberta, Canada: The University of Alberta Press.

Sibley, David Allen. 2000. *The Sibley Guide to Birds.* New York: Alfred A. Knopf.

———. 2000. *The Sibley Guide to Bird Life & Behavior.* New York: Alfred A. Knopf.

Sundstrom, Linea. 1989. *Culture History of the Black Hills with Reference to Adjacent Areas of the Northern Great Plains.* (Reprints in Anthropology, Vol. 40.) Lincoln, NE: J & L Reprint Co.

Tilford, Gregory L., 1997. *Edible and Medicinal Plants of the West.* Missoula: Mountain Press Publishing Co.

Young, Kay. 1993. *Wild Seasons: Gathering and Cooking Wild Plants of the Great Plains.* Lincoln: University of Nebraska Press.

About the Author

LINDA M. HASSELSTROM owns a small family ranch in western South Dakota. Her seventeen published books of poetry and nonfiction include *Feels Like Far: A Rancher's Life on the Great Plains*, autobiographical essays.

Cultural anthropologist Richard Nelson, author of *Patriotism and the American Land,* said of *Feels Like Far:*

> "Linda Hasselstrom knows the land, feels the land, breathes the land, as only a child of the land can do. Her heart was carved by the South Dakota wind. Her bones were made from the South Dakota soil. When Linda Hasselstrom writes, it is the South Dakota prairies writing their own stories. . . . (her) writing is like a black Dakota night riven by lightning flashes and bursts of rain, full of fury and power and raw, brilliant beauty."

With the Great Plains Native Plant Society (www.gpnps.org), Hasselstrom dedicated the Claude A. Barr Memorial Great Plains Garden in 2001 to preserve native shortgrass prairie plants on 350 acres of her ranch, and the Rocky Mountain Bird Observatory, now called the Bird Conservancy of the Rockies (www.birdconservancy.org), established a riparian protection area on her land along Battle Creek.